Synthetic Data for Machine Learning

Revolutionize your approach to machine learning with this comprehensive conceptual guide

Abdulrahman Kerim

BIRMINGHAM—MUMBAI

Synthetic Data for Machine Learning

Group Product Manager: Ali Abidi

Publishing Product Managers: Dhruv J. Kataria and Anant Jain

Senior Editor: David Sugarman

Content Development Editor: Shreya Moharir

Technical Editor: Devanshi Ayare

Copy Editor: Safis Editing

Project Coordinator: Farheen Fathima

Proofreader: Safis Editing

Indexer: Subalakshmi Govindhan

Production Designer: Jyoti Kadam

DevRel Marketing Coordinator: Nivedita Singh

First published: October 2023

Production reference: 1280923

Published by Packt Publishing Pvt. Ltd.
Grosvenor House
11 St Paul's Square
Birmingham
B3 1RB

ISBN 978-1-80324-540-9

www.packtpub.com

To my wife, Emeni, and daughter, Taj.

Contributors

About the author

Abdulrahman Kerim is a full-time lecturer at the University for the Creative Arts (UCA) and an active researcher at the School of Computing and Communications at Lancaster University, UK. Kerim got his MSc in computer engineering, which focused on developing a simulator for computer vision problems. In 2020, Kerim started his PhD to investigate synthetic data advantages and potentials. His research on developing novel synthetic-aware computer vision models has been recognized internationally. He has published many papers on the usability of synthetic data at top-tier conferences and in journals such as BMVC and IMAVIS. He currently works with researchers from Google and Microsoft to overcome real-data issues, specifically for video stabilization and semantic segmentation tasks.

I would like to extend my sincere gratitude to two extraordinary individuals who have been my pillars of strength and support throughout this journey – my loving wife, Emeni, and our precious daughter, Taj.

Emeni, your boundless patience, encouragement, and understanding have been my guiding light. Your belief in my passion for writing has given me the courage to embark on this special journey. Your sacrifices and tireless efforts to create a nurturing home environment for our family have always allowed me to pursue my dreams. You are my muse and my anchor, and I am grateful for your presence in my life.

Taj, even though you may be too young to fully comprehend the significance of this endeavor, your smiles, laughter, and innocent wonder have provided me with a constant source of inspiration. Your presence has infused joy and vitality into my creative process, reminding me of the beauty in life's simplest moments.

To both of you, thank you, for your love and belief in me. Your sacrifices, encouragement, and support have been the driving force behind the creation of this book. This achievement is as much yours as it is mine.

About the reviewers

Oishi Deb is a PhD researcher at the University of Oxford, and her research interests include AI, ML, and AI ethics. Prior to starting her PhD, Oishi gained industrial experience at Rolls-Royce, working in software engineering, data science, and ML. Oishi was named 2017 Student of the Year by the president and vice-chancellor of the University of Leicester, based on her academic performance in her undergraduate degree in software and electronics engineering. Oishi was selected for the DeepMind scholarship program, as part of which DeepMind funded her MSc in AI. Oishi is a chair for an ELLIS reading group in deep learning, and she has also served on the program committee as a reviewer for the NeurIPS and ICML conference workshops.

Leandro Soriano Marcolino is a lecturer (assistant professor) at Lancaster University. He obtained his doctorate degree at the University of Southern California (USC), advised by Milind Tambe. He has published papers on AI, robotics, computer vision, and ML in several key conferences, such as AAAI, AAMAS, IJCAI, CVPR, NeurIPS, ICRA, and IROS. He develops new learning and decision-making techniques for autonomous agents, which usually happen at execution time. Leandro has explored several interesting domains, such as multi-agent teamwork, swarm robotics, computer Go, video games, and more recently, computer vision, including the usage of simulators and synthetic data to create or improve state-of-the-art ML models.

Table of Contents

Part 3: Synthetic Data Generation Approaches

6

Leveraging Simulators and Rendering Engines to Generate Synthetic Data   55

7

Exploring Generative Adversarial Networks   69

8

Video Games as a Source of Synthetic Data   81

9

Exploring Diffusion Models for Synthetic Data   89

Part 4: Case Studies and Best Practices

10

11

12

13

Part 5: Current Challenges and Future Perspectives

14

15

Diversity Issues in Synthetic Data   155

16

Photorealism in Computer Vision   163

17

Conclusion   169

Index   173

Other Books You May Enjoy   184

Preface

Machine learning (**ML**) has made our lives far easier. We cannot imagine our world without ML-based products and services. ML models need to be trained on large-scale datasets to perform well. However, collecting and annotating real data is extremely expensive, error-prone, and subject to privacy issues, to name a few disadvantages. Synthetic data is a promising solution to real-data ML-based solutions.

Synthetic Data for Machine Learning is a unique book that will help you master synthetic data, designed to make your learning journey enjoyable. In this book, theory and good practice complement each other to provide leading-edge support!

The book helps you to overcome real data issues and improve your ML models' performance. It provides an overview of the fundamentals of synthetic data generation and discusses the pros and cons of each approach. It reveals the secrets of synthetic data and the best practices to leverage it better.

By the end of this book, you will have mastered synthetic data and increased your chances of becoming a market leader. It will enable you to springboard into a more advanced, cheaper, and higher-quality data source, making you well prepared and ahead of your peers for the next generation of ML!

Who this book is for

If you are an ML practitioner or researcher who wants to overcome data problems in ML, this book is written especially for you! It assumes you have basic knowledge of ML and Python programming (not more!). The book was carefully designed to give you the foremost guidance to master synthetic data for ML. It builds your knowledge gradually from synthetic data concepts and algorithms to applications, study cases, and best practices. The book is one of the pioneer works on the subject, providing leading-edge support for ML engineers, researchers, companies, and decision-makers.

What this book covers

Chapter 1, Machine Learning and the Need for Data, introduces you to ML. You will understand the main difference between non-learning- and learning-based solutions. Then, the chapter explains why deep learning models often achieve state-of-the-art results. Following this, it gives you a brief idea of how the training process is done and why large-scale training data is needed in ML.

Chapter 2, Annotating Real Data, explains why ML models need annotated data. You will understand why the annotation process is expensive, error-prone, and biased. At the same time, you will be introduced to the annotation process for a number of ML tasks, such as image classification, semantic segmentation, and instance segmentation. You will explore the main annotation problems. At the same time, you will understand why ideal ground truth generation is impossible or extremely difficult for some tasks, such as optical flow estimation and depth estimation.

Chapter 3, Privacy Issues in Real Data, highlights the main privacy issues with real data. It explains why privacy is preventing us from using large-scale real data for ML in certain fields such as healthcare and finance. It demonstrates the current approaches for mitigating these privacy issues in practice. Furthermore, you will have a brief introduction to privacy-preserving ML.

Chapter 4, An Introduction to Synthetic Data, defines synthetic data. It gives a brief history of the evolution of synthetic data. Then, it introduces you to the main types of synthetic data and the basic data augmentation approaches and techniques.

Chapter 5, Synthetic Data as a Solution, highlights the main advantages of synthetic data. In this chapter, you will learn why synthetic data is a promising solution for privacy issues. At the same time, you will understand how synthetic data generation approaches can be configured to cover rare scenarios that are extremely difficult and expensive to capture in the real world.

Chapter 6, Leveraging Simulators and Rendering Engines to Generate Synthetic Data, introduces a well-known method for synthetic data generation using simulators and rendering engines. It describes the main pipeline for creating a simulator and generating automatically annotated synthetic data. Following this, it highlights the challenges and the state-of-the-art research in this field, and briefly discusses two simulators for synthetic data generation.

Chapter 7, Exploring Generative Adversarial Networks, introduces **Generative Adversarial Networks (GANs)** and discusses the evolution of this method. It explains the typical architecture of a GAN. After this, the chapter illustrates the training process. It highlights some great applications of GANs including generating images and text-to-image translation. It also describes a few variations of GANs: conditional GAN, CycleGAN, CTGAN, WGAN, WGAN-GP, and f-GAN. Furthermore, the chapter is supported by a real-life case study and a discussion of the state-of-the-art research in this field.

Chapter 8, Video Games as a Source of Synthetic Data, explains why to use video games for synthetic data generation. It highlights the great advancement in this sector. It discusses the current research in this direction. At the same time, it features challenges and promises toward utilizing this approach for synthetic data generation.

Chapter 9, Exploring Diffusion Models for Synthetic Data, introduces you to diffusion models and highlights the pros and cons of this synthetic data generation approach. It casts light on opportunities and challenges. The chapter is enriched by a discussion of ethical issues and concerns around utilizing this synthetic data approach in practice. In addition to that, the chapter is enriched with a review of the state-of-the-art research on this topic.

Chapter 10, Case Study 1 – Computer Vision, introduces you to a multitude of industrial applications of computer vision. You will discover some of the key problems that were successfully solved using computer vision. In parallel to this, you will grasp the major issues with traditional computer vision solutions. Additionally, you will explore and comprehend thought-provoking examples of using synthetic data to improve computer vision solutions in practice.

Chapter 11, Case Study 2 – Natural Language Processing, introduces you to a different field where synthetic data is a key player. It highlights why **Natural Language Processing (NLP)** models require large-scale training data to converge. It shows examples of utilizing synthetic data in the field of NLP. It explains the pros and cons of real-data-based approaches. At the same time, it shows why synthetic data is the future of NLP. It supports this discussion by bringing up examples from research and industry fields.

Chapter 12, Case Study 3 – Predictive Analytics, introduces predictive analytics as another area where synthetic data has been used recently. It highlights the disadvantages of real-data-based solutions. It supports the discussion by providing examples from the industry. Following this, it sheds light on the benefits of employing synthetic data in the predictive analytics domain.

Chapter 13, Best Practices for Applying Synthetic Data, explains some fundamental domain-specific issues limiting the usability of synthetic data. It gives general comments on issues that can be seen frequently when generating and utilizing synthetic data. Then, it introduces a set of good practices that improve the usability of synthetic data in practice.

Chapter 14, Synthetic-to-Real Domain Adaptation, introduces you to a well-known issue limiting the usability of synthetic data called the domain gap problem. It represents various approaches to bridge this gap. At the same time, it shows current state-of-the-art research for synthetic-to-real domain adaptation. Then, it represents the challenges and issues in this context.

Chapter 15, Diversity Issues in Synthetic Data, introduces you to another well-known issue in the field of synthetic data, which is generating diverse synthetic datasets. It discusses different approaches to ensure high diversity even with large-scale datasets. Then, it highlights some issues and challenges in achieving diversity for synthetic data.

Chapter 16, Photorealism in Computer Vision, explains the need for photo-realistic synthetic data in computer vision. It highlights the main approaches toward photorealism, its main challenges, and its limitations. Although the chapter focuses on computer vision, the discussion can be generalized to other domains such as healthcare, robotics, and NLP.

Chapter 17, Conclusion, summarizes the book from a high-level view. It reminds you about the problems with real-data-based ML solutions. Then, it recaps the benefits of synthetic data-based solutions, challenges, and future perspectives.

To get the most out of this book

You will need a version of PyCharm installed on your computer – the latest version, if possible. All code examples have been tested using Python 3.8 and PyCharm 2023.1 (Professional Edition) on Ubuntu 20.04.2 LTS. However, they should work with future version releases, too.

Software/hardware covered in the book	Operating system requirements
Python 3.8+	Windows, macOS, or Linux
PyCharm 2023.1	

If you are using the digital version of this book, we advise you to type the code yourself or access the code from the book's GitHub repository (a link is available in the next section). Doing so will help you avoid any potential errors related to the copying and pasting of code.

Download the example code files

You can download the example code files for this book from GitHub at `https://github.com/PacktPublishing/Synthetic-Data-for-Machine-Learning`. If there's an update to the code, it will be updated in the GitHub repository.

We also have other code bundles from our rich catalog of books and videos available at `https://github.com/PacktPublishing/`. Check them out!

Conventions used

There are a number of text conventions used throughout this book.

`Code in text`: Indicates code words in text, database table names, folder names, filenames, file extensions, pathnames, dummy URLs, user input, and Twitter (now, X) handles. Here is an example: "Please note that the `seed` parameter will help us to get diverse images in this example."

A block of code is set as follows:

```
//Example of Non-learning AI (My AI Doctor!)
Patient.age //get the patient's age
Patient. temperature //get the patient's temperature
Patient.night_sweats //get if the patient has night sweats
Paitent.Cough //get if the patient coughs
```

> **Tips or important notes**
> Appear like this.

Get in touch

Feedback from our readers is always welcome.

General feedback: If you have questions about any aspect of this book, email us at `customercare@packtpub.com` and mention the book title in the subject of your message.

Errata: Although we have taken every care to ensure the accuracy of our content, mistakes do happen. If you have found a mistake in this book, we would be grateful if you would report this to us. Please visit `www.packtpub.com/support/errata` and fill in the form.

Piracy: If you come across any illegal copies of our works in any form on the internet, we would be grateful if you would provide us with the location address or website name. Please contact us at `copyright@packtpub.com` with a link to the material.

If you are interested in becoming an author: If there is a topic that you have expertise in and you are interested in either writing or contributing to a book, please visit `authors.packtpub.com`.

Share Your Thoughts

Once you've read *Synthetic Data for Machine Learning*, we'd love to hear your thoughts! Scan the QR code below to go straight to the Amazon review page for this book and share your feedback.

`https://packt.link/r/1-803-24540-9`

Your review is important to us and the tech community and will help us make sure we're delivering excellent quality content.

Download a free PDF copy of this book

Thanks for purchasing this book!

Do you like to read on the go but are unable to carry your print books everywhere?

Is your eBook purchase not compatible with the device of your choice?

Don't worry, now with every Packt book you get a DRM-free PDF version of that book at no cost.

Read anywhere, any place, on any device. Search, copy, and paste code from your favorite technical books directly into your application.

The perks don't stop there, you can get exclusive access to discounts, newsletters, and great free content in your inbox daily

Follow these simple steps to get the benefits:

1. Scan the QR code or visit the link below

https://packt.link/free-ebook/9781803245409

2. Submit your proof of purchase

3. That's it! We'll send your free PDF and other benefits to your email directly

Part 1:
Real Data Issues, Limitations, and Challenges

In this part, you will embark on a comprehensive journey into **Machine Learning** (**ML**). You will learn why ML is so powerful. The training process and the need for large-scale annotated data will be explored. You will investigate the main issues with annotating real data and learn why the annotation process is expensive, error-prone, and biased. Following this, you will delve into privacy issues in ML and privacy-preserving ML solutions.

This part has the following chapters:

- *Chapter 1, Machine Learning and the Need for Data*
- *Chapter 2, Annotating Real Data*
- *Chapter 3, Privacy Issues in Real Data*

1

Machine Learning and the Need for Data

Machine learning (**ML**) is the crown jewel of **artificial intelligence** (**AI**) and has changed our lives forever. We cannot imagine our daily lives without ML tools and services such as Siri, Tesla, and others.

In this chapter, you will be introduced to ML. You will understand the main differences between non-learning and learning-based solutions. Then, you will see why **deep learning** (**DL**) models often achieve state-of-the-art results. Following this, you will get a brief introduction to how the training process is done and why large-scale training data is needed in ML.

In this chapter, we're going to cover the following main topics:

- AI, ML, and DL
- Why are ML and DL so powerful?
- Training ML models

Technical requirements

Any code used in this chapter will be available in the corresponding chapter folder in this book's GitHub repository: `https://github.com/PacktPublishing/Synthetic-Data-for-Machine-Learning`.

We will be using **PyTorch**, which is a powerful ML framework developed by Meta AI.

Artificial intelligence, machine learning, and deep learning

In this section, we learn what exactly ML is. We will learn to differentiate between learning and non-learning AI. However, before that, we'll introduce ourselves to AI, ML, and DL.

Artificial intelligence (AI)

There are different definitions of AI. However, one of the best is John McCarthy's definition. McCarthy was the first to coin the term *artificial intelligence* in one of his proposals for the 1956 Dartmouth Conference. He defined the outlines of this field by many major contributions such as the Lisp programming language, utility computing, and timesharing. According to the father of AI in *What is Artificial Intelligence?* (https://www-formal.stanford.edu/jmc/whatisai.pdf):

> *It is the science and engineering of making intelligent machines, especially intelligent computer programs. It is related to the similar task of using computers to understand human intelligence, but AI does not have to confine itself to methods that are biologically observable.*

AI is about making computers, programs, machines, or others mimic or imitate human intelligence. As humans, we perceive the world, which is a very complex task, and we reason, generalize, plan, and interact with our surroundings. Although it is fascinating to master these tasks within just a few years of our childhood, the most interesting aspect of our intelligence is the ability to improve the learning process and optimize performance through experience!

Unfortunately, we still barely scratch the surface of knowing about our own brains, intelligence, and other associated functionalities such as vision and reasoning. Thus, the trek of creating "intelligent" machines has just started relatively recently in civilization and written history. One of the most flourishing directions of AI has been learning-based AI.

AI can be seen as an umbrella that covers two types of intelligence: learning and non-learning AI. It is important to distinguish between AI that improves with experience and one that does not!

For example, let's say you want to use AI to improve the accuracy of a physician identifying a certain disease, given a set of symptoms. You can create a simple recommendation system based on some generic cases by asking domain experts (senior physicians). The pseudocode for such a system is shown in the following code block:

```
//Example of Non-learning AI (My AI Doctor!)
Patient.age //get the patient age
Patient. temperature //get the patient temperature
Patient.night_sweats //get if the patient has night sweats
Paitent.Cough //get if the patient cough
```

```
// AI program starts
if Patient.age > 70:
    if Patient.temperature > 39 and Paitent.Cough:
        print("Recommend Disease A")
        return
elif Patient.age < 10:
    if Patient.tempreture > 37 and not Paitent.Cough:
        if Patient.night_sweats:
                print("Recommend Disease B")
                return
else:
    print("I cannot resolve this case!")
    return
```

This program mimics how a physician may reason for a similar scenario. Using simple `if-else` statements with few lines of code, we can bring "intelligence" to our program.

> **Important note**
>
> This is an example of non-learning-based AI. As you may expect, the program will not evolve with experience. In other words, the logic will not improve with more patients, though the program still represents a clear form of AI.

In this section, we learned about AI and explored how to distinguish between learning and non-learning-based AI. In the next section, we will look at ML.

Machine learning (ML)

ML is a subset of AI. The key idea of ML is to enable computer programs to learn from experience. The aim is to allow programs to learn without the need to dictate the rules by humans. In the example of the AI doctor we saw in the previous section, the main issue is creating the rules. This process is extremely difficult, time-consuming, and error-prone. For the program to work properly, you would need to ask experienced/senior physicians to express the logic they usually use to handle similar patients. In other scenarios, we do not know exactly what the rules are and what mechanisms are involved in the process, such as object recognition and object tracking.

ML comes as a solution to learning the rules that control the process by exploring special training data collected for this task (see *Figure 1.1*):

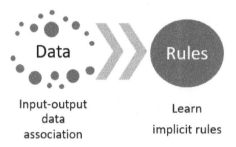

Figure 1.1 – ML learns implicit rules from data

ML has three major types: **supervised**, **unsupervised**, and **reinforcement learning**. The main difference between them comes from the nature of the training data used and the learning process itself. This is usually related to the problem and the available training data.

Deep learning (DL)

DL is a subset of ML, and it can be seen as the heart of ML (see *Figure 1.2*). Most of the amazing applications of ML are possible because of DL. DL learns and discovers complex patterns and structures in the training data that are usually hard to do using other ML approaches, such as **decision trees**. DL learns by using **artificial neural networks** (**ANNs**) composed of multiple layers or too many layers (an order of 10 or more), inspired by the human brain; hence the *neural* in the name. It has three types of layers: input, output, and hidden. The input layer receives the input, while the output layer gives the prediction of the ANN. The hidden layers are responsible for discovering the hidden patterns in the training data. Generally, each layer (from the input to the output layers) learns a more abstract representation of the data, given the output of the previous layer. The more hidden layers your ANN has, the more complex and non-linear the ANN will be. Thus, ANNs will have more freedom to better approximate the relationship between the input and output or to learn your training data. For example, AlexNet is composed of 8 layers, VGGNet is composed of 16 to 19 layers, and ResNet-50 is composed of 50 layers:

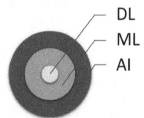

Figure 1.2 – How DL, ML, and AI are related

The main issue with DL is that it requires a large-scale training dataset to converge because we usually have a tremendous number of parameters (weights) to tweak to minimize the loss. In ML, loss is a way to penalize wrong predictions. At the same time, it is an indication of how well the model is learning the training data. Collecting and annotating such large datasets is extremely hard and expensive.

Nowadays, using synthetic data as an alternative or complementary to real data is a hot topic. It is a trending topic in research and industry. Many companies such as Google (Google's Waymo utilizes synthetic data to train autonomous cars) and Microsoft (they use synthetic data to handle privacy issues with sensitive data) started recently to invest in using synthetic data to train next-generation ML models.

Why are ML and DL so powerful?

Although most AI fields are flourishing and gaining more attention recently, ML and DL have been the most influential fields of AI. This is because of several factors that make them distinctly a better solution in terms of accuracy, performance, and applicability. In this section, we are going to look at some of these essential factors.

Feature engineering

In traditional AI, it is compulsory to design the features *manually* for the task. This process is extremely difficult, time-consuming, and task/problem-dependent. If you want to write a program, say to recognize car wheels, you probably need to use some filters to extract edges and corners. Then, you need to utilize these extracted features to identify the target object. As you may anticipate, it is not always easy to know what features to select or ignore. Imagine developing an AI-based solution to predict if a patient has COVID-19 based on a set of symptoms at the early beginning of the pandemic. At that time, human experts did not know how to answer such questions. ML and DL can solve such problems.

DL models learn to *automatically* extract useful features by learning hidden patterns, structures, and associations in the training data. A **loss** is used to guide the learning process and help the model achieve the objectives of the training process. However, for the model to converge, it needs to be exposed to sufficiently diverse training data.

Transfer across tasks

One strong advantage of DL is that it's more task-independent compared to traditional ML approaches. Transfer learning is an amazing and powerful feature of DL. Instead of training the model from scratch, you can start the training process using a different model trained on a similar task. This is very common in fields such as computer vision and natural language processing. Usually, you have a small dataset of your own target task, and your model would not converge using only this small dataset. Thus, training the model on a dataset close to the domain (or the task) but that's sufficiently more diverse and larger and then fine-tuning on your task-specific dataset gives better results. This idea allows your model to transfer the learning between tasks and domains:

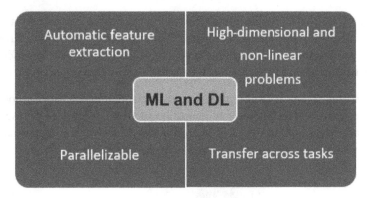

Figure 1.3 – Advantages of ML and DL

> **Important note**
>
> If the problem is simple or a mathematical solution is available, then you probably do not need to use ML! Unfortunately, it is common to see some ML-based solutions proposed for problems where a clear explicit mathematical solution is already available! At the same time, it is not recommended to use ML if a simple rule-based solution works fine for your problem.

Training ML models

Developing an ML model usually requires performing the following essential steps:

1. Collecting data.
2. Annotating data.
3. Designing an ML model.
4. Training the model.
5. Testing the model.

These steps are depicted in the following diagram:

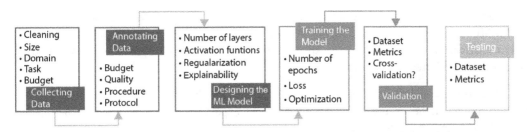

Figure 1.4 – Developing an ML model process

Now, let's look at each of the steps in more detail to better understand how we can develop an ML model.

Collecting and annotating data

The first step in the process of developing an ML model is collecting the needed training data. You need to decide what training data is needed:

- **Train using an existing dataset**: In this case, there's no need to collect training data. Thus, you can skip collecting and annotating data. However, you should make sure that your target task or domain is quite similar to the available dataset(s) you are planning to deploy. Otherwise, your model may train well on this dataset, but it will not perform well when tested on the new task or domain.

- **Train on an existing dataset and fine-tune on a new dataset**: This is the most popular case in today's ML. You can pre-train your model on a large existing dataset and then fine-tune it on the new dataset. Regarding the new dataset, it does not need to be very large as you are already leveraging other existing dataset(s). For the dataset to be collected, you need to identify what the model needs to learn and how you are planning to implement this. After collecting the training data, you will begin the annotation process.

- **Train from scratch on new data**: In some contexts, your task or domain may be far from any available datasets. Thus, you will need to collect large-scale data. Collecting large-scale datasets is not simple. To do this, you need to identify what the model will learn and how you want it to do that. Making any modifications to the plan later may require you to recollect more data or even start the data collection process again from scratch. Following this, you need to decide what ground truth to extract, the budget, and the quality you want.

Next, we'll explore the most essential element of an ML model development process. So, let's learn how to design and train a typical ML model.

Designing and training an ML model

Selecting a suitable ML model for the problem a hand is dependent on the problem itself, any constraints, and the ML engineer. Sometimes, the same problem can be solved by different ML algorithms but in other scenarios, it is compulsory to use a specific ML model. Based on the problem and ML model, data should be collected and annotated.

Each ML algorithm will have a different set of hyperparameters, various designs, and a set of decisions to be made throughout the process. It is recommended that you perform pilot or preliminary experiments to identify the best approach for your problem.

When the design process is finalized, the training process can start. For some ML models, the training process could take minutes, while for others, it could take weeks, months, or more! You may need to perform different training experiments to decide which training hyperparameters you are going to continue with – for example, the number of epochs or optimization techniques. Usually, the loss will be a helpful indication of how well the training process is going. In DL, two losses are used: training and validation loss. The first tells us how well the model is learning the training data, while the latter describes the ability of the model to generalize to new data.

Validating and testing an ML model

In ML, we should differentiate between three different datasets/partitions/sets: training, validation, and testing. The *training* set is used to teach the model about the task and assess how well the model is performing in the training process. The *validation* set is a proxy of the test set and is used to tell us the expected performance of our model on new data. However, the *test* set is the proxy of the actual world – that is, where our model will be tested. This dataset should only be deployed so that we know how the model will perform in practice. Using this dataset to change a hyperparameter or design option is considered cheating because it gives a deceptive understanding of how your model will be performing or generalizing in the real world. In the real world, once your model has been deployed, say for example in industry, you will not be able to tune the model's parameters based on its performance!

Iterations in the ML development process

In practice, developing an ML model will require many iterations between validation and testing and the other stages of the process. It could be that validation or testing results are unsatisfactory and you decide to change some aspects of the data collection, annotation, designing, or training.

Summary

In this chapter, we discussed the terms AI, ML, and DL. We uncovered some advantages of ML and DL. At the same time, we learned the basic steps for developing and training ML models. Finally, we learned why we need large-scale training data.

In the next chapter, we will discover the main issues with annotating large-scale datasets. This will give us a good understanding of why synthetic data is the future of ML!

2
Annotating Real Data

The fuel of the **machine learning** (**ML**) engine is data. Data is available in almost every part of our technology-driven world. ML models usually need to be trained or evaluated on annotated data, not just data! Thus, data by itself is not very useful for ML but annotated data is what ML models need.

In this chapter, we will learn why ML models need annotated data. We will see why the annotation process is expensive, error-prone, and biased. At the same time, you will be introduced to the annotation process for a number of ML tasks, such as **image classification**, **semantic segmentation**, and **instance segmentation**. We will highlight the main annotation problems. At the same time, we will understand why ideal ground truth generation is impossible or extremely difficult for tasks such as **optical flow estimation** and **depth estimation**.

In this chapter, we're going to cover the following main topics:

- The need to annotate real data for ML
- Issues with the annotation process
- Optical flow and depth estimation: ground truth and annotation

Annotating data for ML

In this section, you learn why ML models need annotated data and not simply data! Furthermore, you will be introduced to a diverse set of annotation tools.

Learning from data

As humans, we learn differently from ML models. We just require *implicit* data annotation. However, ML models need *explicit* annotation of the data. For example, let's say you want to train an ML model to classify cat and dog images; you cannot simply feed this model with many images of cats and dogs expecting the model to learn to differentiate between these two classes. Instead, you need to describe what each image is and then you can train your "cat-dog" classifier (see *Figure 2.1*).

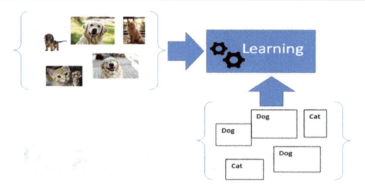

Figure 2.1 – Training data for the cat-dog classifier

It should be noted that the amazing capabilities of ML models are closely related to and highly affected by the quality and quantity of the training data and ground truth. Generally, we need humans to annotate data for two main reasons: training and testing ML models. Next, we will be looking at these in more detail.

Training your ML model

We have four main steps for training an ML model. We will look at each of them next (see *Figure 2.2*).

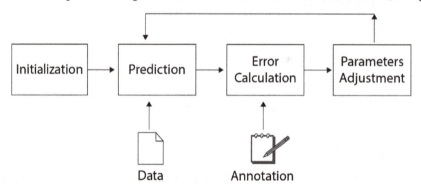

Figure 2.2 – Training process of a typical ML model

Initialization

At the beginning of the training process, the model's parameters should be initialized. Usually, the parameters (weights and biases) of the **deep learning (DL)** model are set to random small numbers because this is what the **stochastic optimization process** expects at the beginning of the optimization process. The stochastic optimization process is a method of finding the best solution for a mathematical problem where randomness and uncertainty are involved to enhance the search procedure.

Prediction

The model utilizes its past knowledge about the task and predicts the output given the input data. We can imagine that the model builds its own understanding of the problem by fitting a **hyperplane** (a decision boundary) to the training data in the training process, and then it projects any given input on this hyperplane to give us the model's prediction for this specific input. Please remember that, at this step, we feed the model with *data only*, without any ground truth. If we go back to our "cat-dog" classifier, the model will be fed with cat and dog images and asked to predict the class or the label of these images.

> **Important note**
>
> *Training*: The ML model develops its own comprehension of the problem by adjusting its parameters and evaluating its performance until it reaches a satisfactory understanding of the training data.
>
> *Testing*: After the training, the ML model is evaluated on new data to assess its performance by using various metrics, such as F1 score, precision, and accuracy.

Error calculation

Given what the model has predicted, we now need to assess the *correctness* of this prediction. This is exactly where we need the ground truth or the annotations. At this step, we have two inputs: the model's prediction and the ground truth. Going back to our "cat-dog" classifier, the model may wrongly predict the class of a cat image as "dog." Now, since we have the true class of the training cat image, we can tell the model that its understanding of the cat-dog problem was wrong this time (for this training sample). Furthermore, we can calculate how close the model was by using the loss function, which depends on the type of the problem. Please note that we essentially have two types of problems in ML: **classification** and **regression** problems. In classification problems, the ML model learns to categorize training data. For example, the "cat-dog" problem is a classification problem, and the error could be 0 or 1 since you have two categories: cat or dog. On the other hand, in regression problems, the ML model learns to leverage input data to predict a continuous value. For example, assume you are training a model to predict a house's price based on some information about the house: location, number of rooms, age, and other information. Assume that the model predicted the house's price to be £100,000 but the actual price (from the ground truth) is £105,000. Then, the error in this case is £5,000.

The error is the essence of the learning process in ML; it provides guidance for training the model – for example, how much it needs to update its parameters and which parameters.

Backpropagation

This is where the learning happens. Given the calculated error, we need to update the model's parameters or weights based on the input and error. In other words, the model needs to "debug" the cause of this error in the prediction. If the error is small, the model will slightly update its weights or understanding of the problem. On the other hand, if the error is huge, the model will need to make major changes to the weights, thus, the understanding of the problem. Going back again to the "cat-dog" classifier, at the beginning of the training process, most predictions will be wrong, thus the model will be updating its weights drastically. In contrast, when the model is close to convergence (the best possible understanding of the training data), ideally, it starts to get most of its predictions right, thus making just slight updates on the weights.

Testing your ML model

To assess the performance of your ML model, you need annotated data too. Thus, the annotations are not just needed for training but also for testing. Usually, qualitative results are good for having an overall understanding of how the model is performing in general or in some individual interesting scenarios. However, quantitative results are the most important way to understand the ML model's robustness, accuracy, and precision.

Using the ground truth, we can examine our trained model's performance on a large number of examples. Thus, there is no need to look at predictions individually as the overall average, standard deviation, and other statistics will be a good description for that. In the next section, we will delve into common issues with the annotation of real data.

Issues with the annotation process

As we have seen so far, annotations are critical to both training and testing. Thus, any mislabeling, biased annotations, or insufficient annotated data will drastically impact the learning and evaluation process of your ML model. As you can expect, the annotation process is time-consuming, expensive, and error-prone, and this is what we will see in this section.

The annotation process is expensive

To train state-of-the-art computer vision or **natural language processing** (**NLP**) models, you need large-scale training data. For example, *BERT* (https://arxiv.org/abs/1810.04805) was trained on BooksCorpos (800 million words) and Wikipedia (2,500 million words). Similarly, *ViT* (https://arxiv.org/abs/2010.11929) was trained on ImageNet (14 million images) and JFT (303 million images). Annotating such huge datasets is extremely difficult and challenging. Furthermore, it is time-consuming and expensive. It should be noted that the time required to annotate a dataset depends on three main elements: the task or problem, dataset size, and granularity level. Next, we will be looking at each of these in more detail.

Task

For example, annotating a dataset for a binary classification problem is easier and requires less time compared to annotating a dataset for semantic segmentation. Thus, the nature of the task also imposes clear difficulty on the annotation process. Even for the same task, let's say semantic segmentation, annotating a single image under standard weather conditions and normal illumination takes approximately 90 minutes for the *Cityscapes* dataset (Marius Cordts, et al. *The cityscapes dataset for semantic urban scene understanding*. In Proceedings of the IEEE conference on computer vision and pattern recognition, pages 3213–3223, 2016). However, doing similar annotation for images under adverse conditions such as snow, rain, and fog or at low illumination such as nighttime takes up to 3 hours for the *ACDC* dataset (Christos Sakaridis, et al. *ACDC: The adverse conditions dataset with correspondences for semantic driving scene understanding*. In Proceedings of the IEEE/CVF International Conference on Computer Vision, pages 10765–10775, 2021.).

Dataset size

As expected, the larger the dataset, the harder it is to annotate. The complexity comes from managing such a huge dataset and ensuring the same annotation and data collection protocol is being followed by a large group of annotators. These annotators may have different languages, backgrounds, experiences, and skills. Indeed, guiding such a huge, diverse team, probably in different geographical locations, is not simple.

Granularity level

The more detail you want your ground truth to capture, the more work for the annotators to perform. Let's take **visual object tracking** as an example. Annotating images for single-object tracking is easier than multi-object tracking. We find the same thing for semantic segmentation, too. Annotating a semantic segmentation dataset with 3 classes is easier than 10 classes. Furthermore, the type of class also creates difficulty for the annotator. In other words, small objects may be harder to differentiate from the background and thus harder to annotate.

Next, we look at the main reasons behind noisy ground truth issues commonly seen in real datasets.

The annotation process is error-prone

In this section, we shed light on the key reasons behind issues in manually annotated real data.

Human factor

The most important element in the annotation process is humans. However, we are limited by our perceptions of the world. Humans struggle to perceive with the naked eye the visual content in scenarios such as low illumination, cluttered scenes, or when objects are far from the camera, transparent, and so on. At the same time, miscommunication and misunderstanding of annotation protocol is another major issue. For example, assume you asked a team of annotators to annotate images for a visual object-tracking training dataset. You aim only to consider the *person object* for this task. Some annotators will annotate humans without objects while other annotators may consider other objects carried by humans as part of the object of interest (see *Figure 2.3*). Furthermore, some annotators may consider only the unoccluded part of the human. This will cause a major inconsistency in the training data and the model will struggle to learn the task and will never converge.

  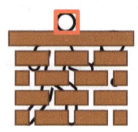

Figure 2.3 – Samples of annotation errors due to unclear annotation protocol

Recording tools

If the recoding camera is shaky, the captured images will be blurred and thus the annotators will fail to accurately identify the actual pixels of the object from the background. Furthermore, the intrinsic and extrinsic parameters of the camera drastically change how the 3D scene will be projected into a 2D image. The focal length of the lens, shutter speed, lens distortion, and others all introduce certain errors in the annotation process. In other words, the objects annotated by annotators may not exactly correspond to the same object in the raw image or even in the 3D world.

Scene attributes

Attributes such as weather conditions and time of the day all play an important role in the annotation process. As we have mentioned earlier, clear weather in the daytime may help the annotators to clearly identify objects as compared to adverse conditions at nighttime. In parallel to this, crowded and cluttered scenes are much more difficult to annotate and more subject to errors (see *Figure 2.4*).

Figure 2.4 – Scene attribute: crowded scenes are more subject to annotation errors

Annotation tools

To enhance the annotation process, there are various annotation tools, such as Labelbox, Scale AI, Dataloop, HiveData, and LabelMe. Some of the annotation tools integrate AI components to optimize the annotation process by assisting the human annotator, such as Labelbox. While these AI-assisted methods are promising, they are not practical and reliable yet. In other words, the human annotator still needs to verify and correct the predictions. Additionally, some of these methods are slow and far from able to provide real-time assistance. In addition to this, if the problem is novel, the AI assistance will not work as expected because the AI model was not trained on similar scenarios.

Given the task, dataset size, and team specifications, a suitable annotation tool should be selected. The annotation tool should be the same for all annotators to ensure consistency among the annotators and the created ground truth.

The annotation process is biased

To understand the world and to reason about it efficiently, our brains build fast decisions and judgments based on our previous experience and systems of beliefs. For more details, see *Decision Making: Factors that Influence Decision Making, Heuristics Used, and Decision Outcomes* (`http://www.inquiriesjournal.com/a?id=180`). ML models learn to reason and perceive the world using the training data. We try to collect and annotate the data objectively. However, unintentionally, we reflect our biases on the data we collect and annotate. Consequently, ML models also become biased and unfair. We will discuss the three common factors of annotation process bias next.

Understanding the problem and task

The annotator may not know the problem, may not understand the data, or understand why the data is collected and annotated. Thus, they may make wrong assumptions or misinterpret data. Furthermore, given the differences between the annotators, they may understand the problem differently.

Background, ideology, and culture

This is a critical factor behind inconsistency in the annotation process. Let's imagine that you asked a group of 10 annotators to annotate a dataset for **action recognition**. You have only two actions: confirmation or negation. Your annotation team members are from the UK, Bulgaria, and India. The Bulgarian annotators will understand and annotate head shaking as "Yes" and nodding as "No." The other annotators will do the opposite. Thus, you will have wrong training data and your model will not learn this task. There are also other scenarios where the bias is not clear and cannot be easily identified, and this is the hardest issue under this scope.

Subjectivity and emotions

For some problems such as **text sentiment analysis**, a well-known NLP technique used to understand textual data, a human annotator may be biased toward certain political parties, football teams, genders, and ethnicities. Thus, the annotations will be biased to the annotator's point of view as well.

> **Common issues in the annotation process**
> Always avoid the following: using the wrong labeling tool, vague annotation protocol, miscommunication between annotators, adding new labels after starting the annotation process, and modifying the annotation protocol during the process.

Optical flow and depth estimation

In this section, we will look at different ML tasks and the followed procedures to generate their corresponding ground truth.

Ground truth generation for computer vision

Computer vision aims at enabling computers to see using digital images. It is not surprising to know that vision is one of the most complex functionalities performed by our brain. Thus, imitating vision is not simple, and it is rather complex for state-of-the-art computer vision models.

Computer vision tasks include semantic segmentation, instance segmentation, optical flow estimation, depth estimation, normal map estimation, visual object tracking, and many more. Each task has its own unique way of generating the corresponding ground truth. Next, we will see samples of these tasks.

Image classification

The training images for this task usually contain one object, which is the object of interest. The annotation for this task is simply looking at each image and selecting one class or more describing the object in the image.

Semantic and instance segmentation

For semantic and instance segmentation, the annotator needs to assign a class label for each pixel in the image. In other words, the annotator is asked to partition the image into different segments where each segment demonstrates one class for semantic segmentation and one instance for instance segmentation. Please refer to `https://github.com/mrgloom/awesome-semantic-segmentation` for an exhaustive list of semantic segmentation methods, such as *U-Net*, *DeepLab*, and *D2Det*. For instance segmentation, please check `https://github.com/topics/instance-segmentation`.

Object detection and tracking

In object detection and tracking, the annotator draws a bounding box around each object in the image. Object tracking works using video to track an initially selected object throughout the video. On the other hand, **object detection** works on images to detect objects as required by the task. Please refer to `https://github.com/topics/object-detection` for a list of well-known and state-of-the-art object detection methods and useful Python libraries, such as *YOLOv5*, *Mask R-CNN*, and *OpenMMLab*. For object tracking, please refer to `https://github.com/topics/object-tracking` for a list of models, such as *SiamMask*, *HQTrack*, and *CFNet*.

Now, we will examine two specific tasks in computer vision that are extremely hard to generate ground truth for using a standard approach. This is basically just an example of the limitation of real data.

Optical flow estimation

Optical flow is the relative apparent motion of objects from one frame to another. The motion could be because of objects or camera motion. Optical flow has many key applications in tasks, such as **structure from motion**, **video compression**, and **video stabilization**. Structure from motion is widely used in 3D construction, and navigation and manipulation tasks in robotics, augmented reality, and games. Video compression is essential for video streaming, storage, and transmission; video stabilization, on the other hand, is crucial for timelapse videos, and videos recorded by drones or head-mounted cameras. Thus, optical flow has enormous applications in practice. For a comprehensive list of optical flow methods such as *SKFlow*, *GMFlow*, and *RAFT*, please refer to `https://github.com/hzwer/Awesome-Optical-Flow`.

Please note that it is extremely hard to generate ground truth for optical flow. Some approaches apply complex procedures to achieve this under many assumptions, such as an indoor environment and a limited number of objects and motions.

Depth estimation

Depth estimation is the task of measuring the distance of each pixel in the scene to the camera. It is essential for 3D vision and has many applications, such as 3D scene reconstruction, autonomous cars and navigation, medical imaging, and augmented reality. Usually, there are two major approaches for depth estimation: one uses monocular images and the other is based on stereo images utilizing epipolar geometry. Similar to optical flow, generating ground truth for depth estimation is extremely hard in the real world. Please refer to `https://github.com/topics/depth-estimation` for a list of recent depth estimation methods, such as *AdaBins*, *SC-Depth*, and *Monodepth2*.

> **Important note**
>
> For optical flow and depth estimation, most of the standard datasets used are synthetic datasets. For optical flow, we can recognize synthetic datasets such as *FlyingChairs*, *FlyingThings3D*, and *Kubric* (`https://github.com/google-research/kubric`). For depth estimation, we can mention *Virtual KITTI*, *DENSE* (`https://github.com/uzh-rpg/rpg_e2depth`), and *DrivingStereo* (`https://drivingstereo-dataset.github.io/`).

Summary

In this chapter, we learned why ML models need annotated real data. At the same time, we explored some of the common issues in the annotation process. Our exploration has led us to a deeper understanding of real data collection and annotation issues, such as being a time-consuming process and subject to annotator errors. Additionally, we covered the limitations of real data for tasks such as optical flow and depth estimation. In the next chapter, we will look specifically at the privacy issues with real data.

In the following chapters of the book, we will continue our exciting journey to understand the limitations of real data and the promising solutions of synthetic data.

3
Privacy Issues in Real Data

ML is becoming an essential part of our daily lives due to its varied applications. Thus, there is a growing concern about privacy issues in ML. Datasets and trained ML models may disclose personal and sensitive information, such as political views, biometric data, mental health, sexual orientation, and other private information.

In this chapter, we will learn about privacy issues and why this is a concern in ML. Furthermore, we will provide a brief introduction to privacy-preserving ML.

In this chapter, we're going to cover the following main topics:

- Why is privacy an issue in ML?
- What exactly is the privacy problem in ML?
- Privacy-preserving ML
- Real data challenges and issues

Why is privacy an issue in ML?

As we discussed in the previous chapters, ML models need large-scale training data to converge and train well. The data can be collected from social media, online transactions, surveys, questionaries, or other sources. Thus, the collected data may contain sensitive information that individuals may not want to share with some organizations or individuals. If the data was shared or accessed by others, and thus the identities of the individuals were identified, that may cause them personal abuse, financial issues, or identity theft.

The complexity of a privacy breach in ML is closely related to the following main three factors:

- ML task
- Dataset size
- Regulations

Let's take a closer look.

ML task

The *task* mainly defines the type of training data that we need to collect and annotate. Thus, some ML tasks, such as weather prediction and music generation, may have fewer privacy issues compared to other ML tasks, such as biometric authentication and medical image processing and analysis.

Dataset size

The larger the dataset, the more issues you will have with privacy. If the dataset is a large-scale one, then you cannot store it on one device. Thus, you may need some cloud services and tools to manage your datasets, such as MongoDB, Hadoop, and Cassandra. Consequently, you may have less control over your data. Thus, you need to give more attention to the tools and services that process or manage your data.

Regulations

Many countries have clear and restrictive data protection regulations. For example, the UK has *The Data Protection Act 2018* (`https://www.legislation.gov.uk/ukpga/2018/12/contents/enacted`). At the same time, similar regulations can be found in the EU, such as the *General Data Protection Regulation* (`https://gdpr-info.eu`). Thus, if you are using personal data, you have to consider data protection principles, such as transparency, being used for their specified purposes, and being kept for a specific period.

If your dataset includes sensitive and confidential information, then you are subject to more restrictions. Your dataset may contain information about biometrics, gender, health, and religious and political opinions. Thus, if you repurpose a dataset that you, your organization, or someone else collected, this action may be illegal.

One of the main limitations of utilizing real datasets is privacy issues. Due to this, we can see why synthetic data is a promising solution for these issues. In the next section, we will take a closer look at the privacy problem and its related topics, such as copyright and intellectual property infringement, which are major issues when collecting a new dataset.

What exactly is the privacy problem in ML?

Within the scope of privacy in ML, there are two main concerns. The first is regarding the dataset itself – that is, how to collect it, how to keep it private, and how to prevent unauthorized access to sensitive information. The second is associated with the vulnerability of ML models to reveal the training data, which we will discuss in the next section. For now, let's examine the issues related to dataset privacy in ML.

Copyright and intellectual property infringement

Copyright is a legal term that's used to protect the ownership of intellectual property. It prevents or limits others from using your work without your permission. For example, if you take a photograph, record a video, or write a blog, your work is protected by copyright. Thus, others may not share, reproduce, or distribute your work without permission. Consequently, images, videos, text, or other information we see on the internet may have restrictive copyrights. Therefore, if you want to collect a dataset, you must consider the copyright problem carefully. As we know, there are different approaches to curating a real dataset. As an ML practitioner, you can do the following:

- **Collect data yourself**: You can use a camera, microphone, sensors, questionaries, and other methods to collect a tremendous amount of data.

- **Collect data from the internet**: You can use search engines such as Google, Yahoo, Baidu, and DuckDuckGo to collect your dataset, similar to how the *ImageNet* (`https://www.image-net.org/`) dataset was collected.

- **Collect data from other datasets**: You can combine different datasets to create a new dataset for a specific application – for example, if you are interested in visual object tracking for only one class – say, humans – in a specific scenario – say, adverse conditions. In this case, you can combine different datasets by excluding irrelevant classes and weather conditions. Thus, you create a new dataset for your particular problem.

Taking a photo of someone with our camera or recording their voice using our hardware does not permit us to use their information. At the same time, asking for permission is not a practical solution for large-scale datasets. To avoid similar issues some, datasets, such as **ImageNet**, were initially collected from the internet. Then, it was annotated by human annotators. It should be noted that not every image we see on the internet can be used to build our dataset; some images have copyright licenses that restrict how they can be used. Thus, the availability of data that you need on the web does not necessarily mean that you can leverage that data for your problem. The same copyright issues can be seen for combining different datasets as well.

If your ML model was trained on, say, Van Gogh paintings and learned how to generate artwork, a question arises regarding who owns the rights to the generated images: Van Gogh, ML engineers, or both? In ML, this is a controversial and complex issue. **ChatGPT** (`https://openai.com/blog/chatgpt`) and **Imagen** (*Photorealistic Text-to-Image Diffusion Models with Deep Language Understanding*) are examples of such models. Thus, training your model on a real dataset may not give you the right to fully leverage the potential of your ML model. You may still be subject to copyright and the intellectual properties of the samples used in the training process. As you may expect, leveraging synthetic data seems a promising solution to these problems.

Privacy and reproducibility of experiments

One of the main issues in ML-based research is the difficulty in reproducing the experiments, and thus validating the results claimed in research papers. Data privacy is one of the key factors behind this issue. Many companies develop ML solutions using their own datasets. They may share trained models, but not the dataset itself because of many reasons related to privacy issues and regulations. Consequently, other researchers cannot reproduce the experiments and results. This creates a good chance for errors, bias, misinterpretation, falsification, and manipulation of research results. This is also another reason why using synthetic data in certain fields, such as healthcare and finance, can make the research more transparent and the results more trustworthy.

Privacy issues and bias

Bias is another issue closely related to privacy problems in ML models and real data. ML models can be trained on biased data, which can result in biased outcomes. For example, some face recognition commercial applications were found to be less accurate in recognizing people with darker skin, females, or people aged between 18 and 30. The issue is due to the bias in the training data of the ML models. Thus, when the training data is not available because of privacy concerns, certain bias issues may arise. Consequently, this can lead to unequal treatment and discrimination based on race, gender, and other factors. Next, we will delve into traditional solutions to privacy issues.

Privacy-preserving ML

Privacy-preserving ML is a hot topic since it proposes solutions for privacy issues in the ML field. Privacy-preserving ML proposes methods that allow researchers to use sensitive data to train ML models while withholding sensitive information from being shared or accessed by a third party or revealed by the ML model. In the first subsection, we will examine common methods for mitigating privacy issues in datasets.

Approaches for privacy-preserving datasets

In this section, we'll delve into standard approaches for handling and protecting sensitive information in datasets. We will look at anonymization, centralized data, and differential privacy.

Anonymization

Anonymization can be considered one of the earliest approaches for privacy issues in datasets. Let's assume you are given a dataset of patients' medical records that contains their addresses, phone numbers, and postal codes. To anonymize this dataset, you can simply remove this sensitive data while keeping other medical records. Unfortunately, this approach does not work very well for two reasons. Sometimes, you cannot remove this sensitive information because they are part of the task. At the same time, anonymization may not be sufficient, and thus individuals may be identified by linking and combining other information in the dataset or other datasets. As an example, check the paper titled

The 'Re-Identification' of Governor William Weld's Medical Information: A Critical Re-Examination of Health Data Identification Risks and Privacy Protections, Then and Now (`https://papers.ssrn.com/sol3/papers.cfm?abstract_id=2076397`):

Name	Surname	Has diabetes	Smoking	Gender	Age	Occupation
Mike	Chris	Yes	No	Male	60	Programmer
Emma	Cunningham	Yes	Yes	Female	40	Lawyer
Jan	Wan	No	No	Male	32	PhD Student
Olivia	Cunningham	No	No	Female	23	Student

Figure 3.1 – A sample dataset that includes sensitive information

Centralized data

Another approach is based on the idea of keeping the data in a private server and not sharing the data itself but instead answering queries about the data. For example, here are some examples of queries regarding the dataset shown in *Figure 3.1*:

- Return the minimum age of patients with diabetes

- Return the number of female patients with diabetes

- Return the average age of patients with diabetes

As you can see, organizations can keep sensitive data while still allowing other researchers or organizations to leverage this data. However, this approach is not resilient against cyberattacks. These queries can be used by attackers to identify individuals and disclose sensitive information by combining complex queries and linking information from other datasets. For example, if you remove the **Name** and **Surname** columns but still use information regarding **Gender**, **Age**, and **Occupation**, it may still be possible to use these details to identify some patients.

Differential privacy

As a complementary solution to the previous approaches, differential privacy seems a promising solution. However, there are some limitations, as we will see later.

The key idea of this approach is keeping individuals' information secured while still learning about the phenomena under consideration. This approach is based on queries to a server that contains all the sensitive data. Many companies utilize differential privacy, such as Uber, Google, and Apple. The algorithm may add random noise to the information in the dataset or the queries, as shown in *Figure 3.2*:

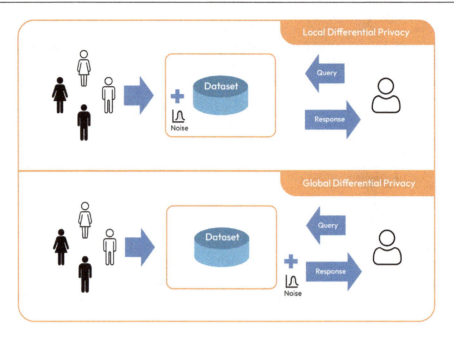

Figure 3.2 – Local and global differential privacy

Differential privacy has two main approaches:

- **Local differential privacy (LDP)**: The data taken from the users are processed with noise, such as Laplacian noise. This process makes the data more secure against attacks. Even if attackers can access the data, they will have a noisy version and they will not be able to identify individuals.

- **Global differential privacy (GDP)**: The raw data taken from clients is not altered but directly stored in the server without any noise being added. When a query is received, the raw answer (accurate) is returned after noise has been added to it. This process generates a private answer to protect data privacy.

Thus, with these data protection mechanisms, it's supposed that ML models can now be trained on sensitive data without disclosing individual sensitive information. Unfortunately, this assumption is not always valid, as we will see in the next few sections.

Approaches for privacy-preserving ML

Privacy-preserving machine learning (PPML) aims to prevent training data leakage. ML models may memorize sensitive information in the training process, and thus confidential information may be revealed by the ML model. Standard PPML methods rely on differential privacy, which we discussed earlier. Next, you will be introduced to federated learning.

Federated learning

Federated learning is a novel strategy that allows ML to be trained on sensitive data while not transferring the data outside of the local server or node. Thus, organizations can share their sensitive data to train ML models while keeping the data in the organization's local servers. At the same time, this is a solution for regulation that prevents sharing data with external parties. It is based on the paradigm of decentralized learning, which we will see next. Please refer to `https://github.com/topics/federated-learning` for a wide range of federated learning frameworks and libraries such as *FATE*, *FedML*, and *PySyft*.

To begin, let's differentiate between centralized and decentralized ML systems:

- **Centralized ML system**: In centralized ML systems, all the training is done in one server. It is easier to implement, and thus it was traditionally applied to ML problems. However, it has many limitations due to the communication between the users and the central server. As shown in *Figure 3.3*, the users need to send information to the central server where the training process will be executed. In addition to the latency issues, this approach is more vulnerable to attacks:

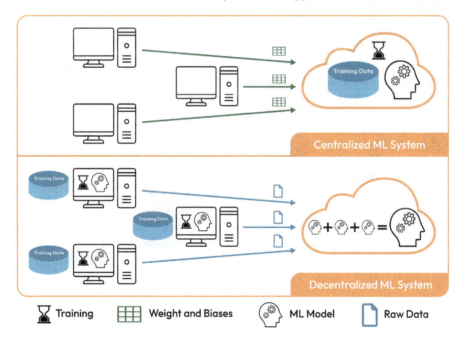

Figure 3.3 – Centralized and decentralized ML systems

- **Decentralized ML system**: In contrast to the centralized option, this system allows the ML model to be trained on the client node rather than the centralized server. The clients send the weights and biases of their trained ML models to the admin server. Following this, the weights and biases are utilized to construct the final ML model. This is a clever solution to many privacy issues. For example, many hospitals in the EU cannot share their patients' data with organizations or people outside the hospital. Thus, using the decentralized ML system, other organizations may access patient data since their data will not leave the hospital's servers and the training process will also be done on their servers.

In general, we can see that decentralized ML systems have the following advantages:

- Less communication between the server and clients

- Better privacy as the clients do not share raw data; instead, they share weights and biases

- Better privacy since data is stored in the local nodes and does not leave the organization during the training process.

In the next section, we will briefly discuss the essence of the privacy issues in real data.

Real data challenges and issues

So far in this chapter, we have presented different approaches for mitigating the privacy issues in real data. As you can see, it is clear that these approaches have limitations and they are not always practical. One fundamental issue is that ML models memorize the training data. Thus, given a trained ML model, it may be possible to retrieve some of the training data. Many researchers recently raised a red flag about the privacy of ML models, even after applying standard privacy solutions. For more information, please refer to *How To Break Anonymity of the Netflix Prize Dataset* (https://arxiv. org/abs/cs/0610105) and *The Secret Sharer: Evaluating and Testing Unintended Memorization in Neural Networks* (https://arxiv.org/abs/1802.08232).

The nature of the real data is the essence of the problem. For instance, if you are given real human faces and you do some operations to anonymize this data or if you apply state-of-the-art approaches for PPML training, the data is still at risk of being divulged. Thus, it seems that instead of proposing more complicated and sometimes impractical solutions for privacy issues in real data, it's time to look at other alternatives and focus on the essence of the problem, which is real data. In other words, it seems that synthetic data represents a rich and safe source for training large-scale ML models, therefore solving these complex privacy issues.

Summary

In this chapter, we discussed why privacy is a critical problem in ML. At the same time, we learned what exactly the privacy problem in this field is. We also learned about the main standard solutions and their limitations.

In the next chapter, we will discover what synthetic data is. This will help us build a solid foundation so that we can learn how to utilize synthetic data as a solution to the real data problems that we examined in the previous chapters.

Part 2:
An Overview of Synthetic Data for Machine Learning

In this part, you will be introduced to synthetic data. You will learn about the history and the main types of synthetic data. Then, you will explore its main advantages. You will understand why synthetic data is a promising solution for many complex issues, such as privacy, that hinder the progress of ML in certain fields. You will learn how synthetic data generation approaches can be utilized to generate data for rare scenarios that are usually expensive and dangerous to capture with real data.

This part has the following chapters:

- *Chapter 4, An Introduction to Synthetic Data*
- *Chapter 5, Synthetic Data as a Solution*

4

An Introduction
to Synthetic Data

In this chapter, we will define and introduce synthetic data. We will briefly explore the history and evolution of synthetic data. Then, we will introduce the main types of synthetic data and the basic data augmentation approaches and techniques.

In this chapter, we're going to cover the following main topics:

- What is synthetic data?
- History of synthetic data
- Synthetic data types
- Data augmentation

Technical requirements

The code used in this chapter will be available under the corresponding chapter folder in the book's GitHub repository: `https://github.com/PacktPublishing/Synthetic-Data-for-Machine-Learning`.

What is synthetic data?

Synthetic data is artificially generated data: the data is not captured, measured, or recorded from the real world. Instead, algorithms or software were used to create or generate this data. Synthetic data can be generated by simulating natural phenomena using mathematical models or by applying some approximations of real-world processes. There are many approaches to generating synthetic data, such as leveraging game engines, such as Unreal and Unity, or utilizing statistical models, such as GANs and diffusion models. As we know, ML models require large-scale training datasets for training and evaluation. Collecting and annotating these datasets is extremely time-consuming, error-prone, and subject to privacy issues. Please refer to *Chapters 2* and *3*. Synthetic data is a powerful solution to address these previous limitations.

Synthetic data is useful for scenarios where collecting and annotating data is expensive, but its applications go beyond this particular use case, as we will see later. Synthetic data has been used in AI, ML, and data analytics, specifically for **computer vision** tasks, which usually require large and hard-to-annotate data for training. Thus, synthetic data has been widely utilized in this field and has shown great progress.

Synthetic data can be generated to train or evaluate ML models under certain conditions that are usually hard to capture in the real world. For example, let us assume we would like to train a computer vision model to predict road traffic accidents based on some visual information such as RGB and LiDAR images. We would need to feed our training model with a sufficiently large dataset that includes thousands of road traffic accidents. Collecting this dataset from the real world may take us weeks, months, or years; it would require many engineers and annotators, and a huge budget to achieve our aim. At the same time, our dataset may not be valid in other countries or after a few years. If you collected your dataset in the UK, where people drive on the left, this dataset would not be applicable to China, where people drive on the right! In parallel to this, if you collected your dataset in 2005, your dataset may not be applicable for 2024 because of, for instance, new car models.

On the other hand, if you generate your synthetic training data using a simulator such as *CARLA* (`https://carla.org`), you can simulate thousands of road traffic accidents. Additionally, you can control car models, scene attributes, weather conditions, and other attributes. This is just an example of the advantages of synthetic data for the training and evaluation of ML models.

Synthetic and real data

Assume that you want to train your ML model to predict the profit of selling a given product. The maximum profit that you can get is £10 and the maximum loss is £10 as well. In the real world, this specific financial problem can be modeled using a simple sinusoidal wave as shown in *Figure 4.1*. Thus, to get the maximum profit, you must sell the product on certain days approximately close to the second and eighth days from the production day (day 0). The blue line gives us the actual model of this problem in the real world. However, this model is hidden from us because if we know this model, then there is no need to use ML for the prediction. Assume that real data perfectly represents this model. Then, the synthetic data in this scenario is the black dots. In other words, synthetic data approximates the real data and real data is also an approximation of the actual process or phenomenon in the real world.

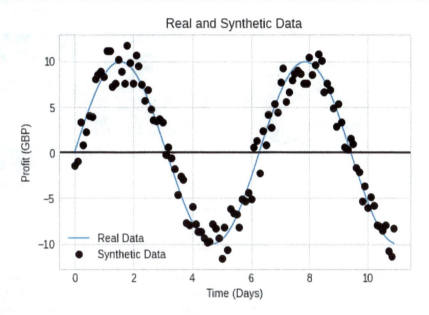

Figure 4.1 – Simple example of real and synthetic data

At this point, we may ask, why do we need to use synthetic data if real data is available? Indeed, this is a sound question: if you have sufficient, annotated, and unbiased real data with no privacy issues, you should not use synthetic data!

Unfortunately, in most real-world problems, such optimal datasets are unavailable, extremely expensive, and limited. Thus, synthetic data comes as a last resort to help ML models learn about the process or the task even when the real data is limited or absent. As you can see in *Figure 4.1*, given only the scatter plot, we can still observe a clear pattern, that is, a **sine wave**, in this data. Thus, a suitable ML model will still be able to learn how to predict what is the best time to sell this product even if you train it using only synthetic data.

Data-centric and architecture-centric approaches in ML

In the field of ML, there are two primary approaches: a model-centric approach, which focuses on the ML model and its architecture, and a data-centric approach, which prioritizes the data as shown in *Figure 4.2*.

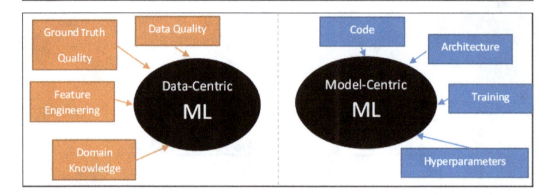

Figure 4.2 – Data-centric and model -centric ML

Next, let us discuss these two approaches in more detail.

- **Model-centric ML**: The model-centric paradigm has been the predominant approach in ML until recently. This approach assumes that the dataset is fixed and strives to come up with a better architecture, novel training procedures, and new ways to search and find optimal hyperparameters. Let us focus on these elements and discuss them in more detail:

  - **Code and architecture**: Researchers continuously develop new architectures to better leverage and learn about the training data. For example, after the release of the well-known *ImageNet Large Scale Visual Recognition Challenge (ILSVRC) dataset* (`https://www.image-net.org/challenges/LSVRC`), many architectures were proposed to improve the top-5 classification error, such as AlexNet, *ImageNet Classification with Deep Convolutional Neural Networks* (`https://proceedings.neurips.cc/paper/2012/file/c399862d3b9d6b76c8436e924a68c45b-Paper.pdf`), and ResNet, *Deep Residual Learning for Image Recognition* (`https://arxiv.org/abs/1512.03385`). The improvements in the ML model can be in the number of layers, that is, the deep network; model parameters; learning filters; and others.

  - **Training**: Training is an interesting research area for ML researchers. Researchers try to find faster ways to train complex ML models and use less data. This addresses issues such as model parameter initialization techniques and their impact on the optimization process, proposes novel optimization techniques, better generalization, less overfitting, novel pretraining techniques, and better fine-tuning tricks for ML models.

  - **Hyperparameters**: Parameters such as the learning rate, batch size, and the number of layers highly impact the overall learning process and thus the model's performance in the real world. Different approaches have been proposed to search for optimal hyperparameters efficiently to further improve ML models.

- **Data-centric ML**: This approach started to get more momentum just recently. It focuses on the data itself rather than the architecture and code. It assumes the ML model's architecture is fixed and strives to improve performance by focusing only on the dataset. It gives more attention to the following data-related concepts:

  - Data quality

  - Ground-truth quality

  - Feature engineering

  - Domain knowledge

Let's see how synthetic data was developed.

History of synthetic data

In this section, we will learn about the evolution of synthetic data. Basically, we can categorize the use of synthetic data into the following categories, which may not reflect the chronological order, as it is very hard to track the early uses of synthetic data for each category.

Random number generators

Random number generators are one of the simplest forms of synthetic data. Assume you are training an ML model to recognize faces. Let us say you have only a limited number of images. You can add, for example, random noise to the original images to create new synthetic ones. The implementation of random noise is possible through the utilization of random number generators. This will help the face recognizer ML model to learn how the person's face changes under certain types of noise (see *Figure 4.3*).

Figure 4.3 – Utilizing random number generators to generate synthetic images

Next, we'll learn about GANs, which are another step in the development of synthetic data.

Generative Adversarial Networks (GANs)

GANs were introduced in 2014 by the famous **NeurIPS** (formerly **NIPS**) paper titled *Generative Adversarial Nets* by Ian Goodfellow et al. (https://proceedings.neurips.cc/paper/2014/file/5ca3e9b122f61f8f06494c97b1afccf3-Paper.pdf). Since then, GANs have been utilized in various applications, such as generating human faces, photo inpainting, 3D object generation, text-to-image translations, and many more interesting applications.

A typical GAN is composed of two networks: a generator and a discriminator. The **generator** receives a noise random input vector and outputs a synthetic sample, for instance, say, a car image. The generator aims at making the synthetically generated data, for example, the car image, indistinguishable from the real data, real car images. The **discriminator**, on the other hand, strives to identify synthetic data from real data. The discriminator is fed with real or synthetic data and asked to predict the data source of the training sample. If the data sample was drawn from real data and the discriminator correctly identified the data source as real data, no error is backpropagated to the discriminator. On the other hand, the generator is penalized for predicting a sample that is distinguishable from the real dataset. Similarly, if the discriminator failed to identify the source of the image, the discriminator is penalized, and the generator is rewarded for generating indistinguishable synthetic samples close to the real dataset (see *Figure 4.4*).

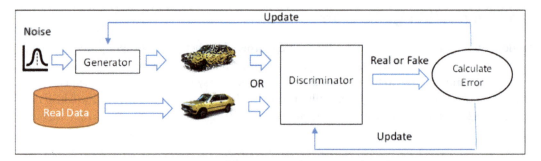

Figure 4.4 – A typical GAN training process

We will discuss GANs in more detail in *Chapter 7*.

Synthetic data for privacy issues

As we discussed in *Chapter 3*, there are enormous privacy issues in real data, and current solutions are only partial solutions to the problem. Recently, synthetic data was proposed as a legitimate solution to these privacy issues. Usually, financial data is often associated with privacy issues as it is problematic to share customers' data, such as personal details, transactions, assets, and income. This information usually is stored in tables. Surprisingly, it was shown that real data can be learned, and synthetic data can be generated. For example, the researchers who authored the paper titled *Modeling Tabular data using Conditional GAN* (`https://arxiv.org/abs/1907.00503`) demonstrated that their **Conditional Tabular GAN (CTGAN)** can model the probability distribution of tabular real data with a complex distribution. Their code can be accessed from the paper's GitHub repository at `https://github.com/sdv-dev/CTGAN`.

Synthetic data in computer vision

Computer vision is one of the main fields in ML that requires large-scale training data. As we discussed earlier, collecting and annotating data for computer vision tasks is extremely expensive and the annotation process is error-prone. As a solution, researchers started to utilize various methods to generate synthetic data such as game engines, video games, GANs, and **Variational Autoencoders (VAEs)**. The huge advancement in game engines such as *Unreal* (`https://www.unrealengine.com`) and *Unity* (`https://unity.com`) facilitated the creation of photorealistic 3D virtual worlds and thus the generation of high-quality and large-scale synthetic data. At the same time, the availability of powerful and affordable **Graphics Processing Units (GPUs)** for small research groups further popularized such game engines.

Synthetic data and ethical considerations

As synthetic data is gaining more attention and being utilized in various applications, in the last few months, many researchers, scientists, artists, and even the public started to question the copyright issues in texts and images generated using models, such as *Chat-GPT* (`https://chat.openai.com/chat`) and *Stable Diffusion* (`https://stablediffusionweb.com`). At the same time, other issues such as accountability and transparency are being brought to light by the ML community for further precautions and more research.

Next, we will dive into the world of synthetic data and learn about its main types in ML.

Synthetic data types

There are various synthetic data types, such as textual, imagery, point cloud, and tabular. Based on the ML problem and task, different types of data are required. In this section, we will discuss the main types of synthetic data in more detail.

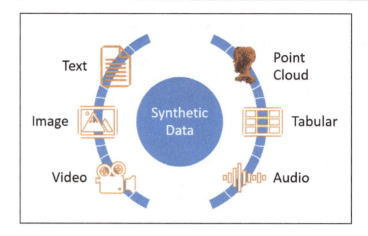

Figure 4.5 – A sample of synthetic data types

- **Text**: Wikipedia, digital books, lexicons, and text corpora are examples of textual data. ML models can be trained on large-scale textual datasets to learn the structure of the text that we generate or write as humans. Then, these models can be leveraged to answer questions, summarize texts, or translate from one language to another. These models, such as *ChatGPT*, *ChatSonic* (`https://writesonic.com`), and *Jasper Chat* (`https://www.jasper.ai`), work by generating synthetic texts based on making predictions on what word should come next.

- **Video, image, and audio**: ML models can learn the patterns in a video, image, or audio, and then they can generate synthetic ones with some new conditions. Models such as *Stable Diffusion* (`https://github.com/CompVis/stable-diffusion#stable-diffusion-v1`), *DALL·E 2* (`https://openai.com/dall-e-2`), and *Imagen* (`https://imagen.research.google`) can be leveraged to generate, theoretically, an unlimited number of synthetic images under various conditions.

- **Tabular**: This refers to data that is usually organized in rows and columns using tables. Typically, rows are the observations and columns are the attributes. ML models can be used to predict missing values in tabular data, for example, *Diffusion models for missing value imputation in tabular data* (`https://arxiv.org/abs/2210.17128`).

- **Point cloud**: This describes the data that captures the 3D coordinates of an object or a space. This data is often collected using photogrammetry or laser scanners. However, the process of capturing the data in the real world is noisy and problematic. Thus, synthetic datasets are seen as a promising solution in this area. As an example, the `SynthCity` dataset was proposed in this paper titled *SynthCity: A large scale synthetic point cloud* (`https://arxiv.org/pdf/1907.04758.pdf`), which provides more than 360 million synthetic point clouds.

In the next section, we delve into data augmentation techniques in ML.

Data augmentation

Data augmentation is a simple yet powerful tool to mitigate overfitting problems, particularly when limited real data is available. Data augmentation techniques aim to leverage domain knowledge to enrich the available training data. Thus, data augmentation is usually applied only to the training data and not to validation or test data. For example, assume you are training a face recognition algorithm and you have only 10 images per person. We can simply double the number of these training samples if we horizontally flip the images. Furthermore, we can enhance the diversity of our training data by applying various transformations, such as shifting, scaling, and rotating, using random variables. Instead of using fixed values for these transformations, we can leverage a random number generator to generate new values for each training epoch. Thus, the ML model will be exposed to new variations of our training data at each training epoch. This simple data augmentation technique will help the model in the training process. There are various data augmentation techniques for images, audio, and texts. Next, let us discuss some of these techniques. Please refer to *Image Data Augmentation for Deep Learning: A Survey* (https://arxiv.org/abs/2204.08610) for more details and techniques for image data augmentation.

Geometric transformations

When limited training images are available and acquiring new ones is expensive, we can apply geometric transformations to the original images, such as translation, rotation, cropping, and flipping. However, it is important to take care that the semantic meaning of the image is preserved after these operations. For example, for cats-versus-dogs classification training images, flipping the image horizontally is acceptable, but a vertical flip is not. Similarly, horizontal and vertical flips may not be valid for traffic sign recognition tasks (see *Figure 4.6*).

Figure 4.6 – Sample of valid and invalid geometric transformations

Translation is simply shifting an image horizontally or vertically by a fixed or random number of units to avoid object bias. For example, assume all cat images in your cats-dogs classification dataset are in the upper right of the image. Then, the ML model will develop a wrong association between the cat class and the upper right of the image. **Rotation** refers to rotating the image at a specific angle clockwise or anticlockwise. Like flipping, for some applications, a specific range may be valid but other ranges may change the semantic meaning of the training image. **Cropping** is cutting the image using a virtual cropping window. It is possible to use a fixed or dynamic cropping window size (height and width).

Noise injection

This technique can be applied to almost all data types and specifically to audio and images. Noise can be drawn from various probability distributions, such as normal (Gaussian), uniform, Poisson, and Bernoulli. As expected, training an ML model with carefully augmented data makes the model more robust against similar noise types. The injected noise can be utilized to simulate issues in a camera lens, microphone, transmission medium, and other sorts of distortions. When the ML model learns how to deal with similar scenarios in the training process, it will not struggle when these scenarios occur in the real world due to unpredictable factors, such as adverse weather conditions and hardware failures or other issues.

Text replacement, deletion, and injection

These techniques are widely used to increase the size of the textual datasets when training a **Natural Language Processing (NLP)** model. It should be noted that like other data augmentation techniques for images, we should pay attention that augmenting the text does not change the meaning of the sentence. **Text replacement** is a simple technique that can be used to generate synthetic text to further enrich the diversity of the training data. A basic augmentation pipeline is shown in *Figure 4.7*. Assume we have the sentence "Synthetic Data is essential in ML" and we want to apply a text augmentation technique to it. Given the sentence, we can randomly select a word, in this example, "essential," and replace it with one of its synonyms selected at random, for example, "crucial." The augmented synthetic sentence becomes "Synthetic Data is crucial in ML."

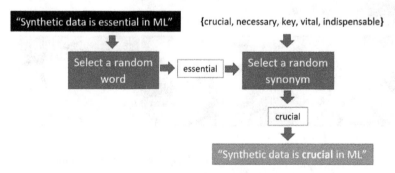

Figure 4.7 – Text augmentation pipeline using synonyms

Similarly, **text deletion** and **text injection** can be utilized to generate synthetic text to improve the performance of ML models.

Summary

In this chapter, we explored synthetic data and its evolution. We learned about the main types of synthetic data. In this chapter, we also discussed the key data augmentation techniques to enrich a limited real dataset for images, audio, and textual data.

In the next chapter, we will bring to light how synthetic data is being used as a solution for problems such as privacy and data scarcity. Additionally, we will learn why it is better in terms of cost and why it is a revolutionary solution for rare and limited real data.

5
Synthetic Data as a Solution

This chapter highlights the main advantages of synthetic data. You will learn why synthetic data is a promising solution for privacy issues. At the same time, you will understand how synthetic data generation approaches can be configured to cover rare scenarios that are extremely difficult and expensive to capture in the real world.

In this chapter, we're going to cover the following main topics:

- Synthetic data generation methods
- The main advantages of synthetic data
- Synthetic data as a revolutionary solution for privacy issues
- Synthetic data as a revolutionary solution for cost and time efficiency issues
- Synthetic data as a revolutionary solution for rare data

The main advantages of synthetic data

As we have seen so far, synthetic data has a wide set of applications because of its enormous advantages. Let's highlight some of these advantages:

- Unbiased
- Diversity
- Data controllability
- Scalable
- Automatic data generation
- Automatic data labeling
- Annotation quality
- Low cost

Figure 5.1 highlights some of the key benefits:

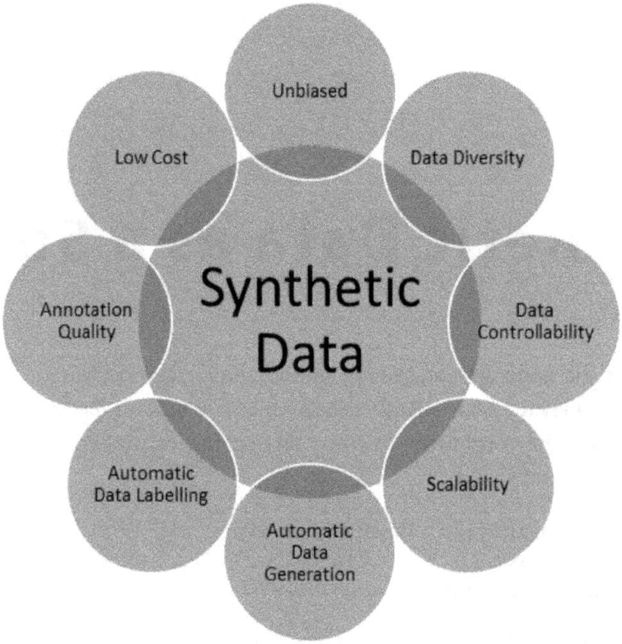

Figure 5.1 – The main advantages of synthetic data

Next, we will delve into each of these advantages. We will see the limitations of real data and how synthetic data is a solution.

Unbiased

Real data is curated and annotated by human annotators. In practice, it is easy for humans, intentionally or accidentally, to neglect or overemphasize certain groups in the population based on some attributes, such as ethnicity, skin color, gender, age, or political views. This creates a biased dataset that negatively affects both training and testing ML models since biased training data gives a corrupted representation of the studied phenomenon or processes that occur in the real world. As a consequence, the ML model will be biased in its decisions. This bias may lead to race, sex, or age discrimination, which causes tremendous unwanted consequences on companies' reputations, customers, and revenue.

Let's discuss one example in NLP. Researchers in the paper titled *The risk of racial bias in hate speech detection* (`https://aclanthology.org/P19-1163.pdf`) demonstrated that tweets by African Americans are two times more likely to be flagged as offensive by automatic hate speech detection ML models. They link the reason to biased training data because of annotators' bias.

Another example comes from computer vision with racial discrimination in face recognition ML algorithms. Face recognition ML models developed by Microsoft, IBM, and Amazon were shown to be less accurate at predicting certain genders and skin colors. These ML models were specifically identified to be less accurate at predicting darker female faces. Some of the largest tech companies in the world, such as Microsoft and IBM, proposed immediate actions to improve their data collection process to mitigate the bias problem in their ML models. Please refer to *Racial Discrimination in Face Recognition Technology* (`https://sitn.hms.harvard.edu/flash/2020/racial-discrimination-in-face-recognition-technology`) for a thorough discussion and more details.

As you might expect, synthetic data can be automatically generated and annotated. Thus, the error by human element factor can be removed or minimized in data generation and annotation processes. Therefore, fewer human errors are expected with synthetic data. At the same time, data can be generated so that it's evenly distributed over the population. Thus, unbiased training and testing data can be generated to support various applications and scenarios. Additionally, in the case of any issues with dataset bias, synthetic data generation approaches can easily be reconfigured to address these problems, which is much faster compared to real data.

Diverse

The process of generating synthetic data can be customized to cover rare cases and scenarios that are not frequent, not easy to capture, or too expensive to annotate. Although it is possible to curate and annotate a diverse real dataset, it is extremely hard to achieve and requires more effort, time, and budget. For instance, capturing real data under adverse weather conditions is harder compared to normal weather. At the same time, capturing data is extremely hard during natural disasters and catastrophes, such as earthquakes, wildfires, hurricanes, tornados, and volcanoes. This limits the usability of ML models under similar scenarios. Therefore, **Life-Critical Systems** or **Safety-Critical Systems (SCSs)** that are based on ML models may fail or malfunction under these scenarios. This failure may cause death, injuries, damage to equipment, and harm to the environment. In addition to this, there are instances where certain events occur frequently, yet they are challenging to capture, such as burglary, sexual exploitation of children, domestic abuse, scams and fraud, street harassment, and terrorism.

Synthetic data generation techniques such as statistical models or simulators can be designed or configured to cover all these scenarios. Thus, it makes the ML models more robust against similar scenarios, which saves lives and properties and also protects societies. Additionally, it opens the door for researchers to focus specifically on rare conditions, events, and scenarios. As a result, diverse and balanced synthetic datasets can be generated to advance research in various fields. At the same time, synthetic data can augment the available real datasets so that they become more diverse and well-balanced.

Controllable

Nature and its entangled processes and phenomena generate real data. For example, you can collect images of human faces to train a face recognition algorithm, but you will have no control over the data generation process itself. You may choose to consider or neglect certain attributes but you cannot perturb the process of how a human face may look in the real world! You can use filters or any image processing technique you want but still, you are only changing how you perceive these images. In other words, you do not make the world desaturated when you wear sunglasses: you only perceive the world as being darker with reduced color saturation. These are two completely different ideas!

Synthetic data allows you to change the world itself as you want. For instance, you can create a city where all people wear the same clothes, walk the same way, and hold the same items! A good question you may ask here is, *why would ML researchers or practitioners want to do this in the first place?*

Being able to control the process of synthetic data generation is extremely important in the field of ML. It allows you to train and test your model on extremely rare conditions, analyze the algorithm's robustness and accuracy, and reiterate the assumptions, the ML model's design, and the training data. Synthetic data gives you the ability to control the parameters of the environment, the elements, and their interactions with each other. Consequently, there is no need to generate irrelevant or redundant data that does not help your ML model. Thus, with synthetic data, you can manage, control, and guide the data generation and annotation processes to achieve your objectives.

Scalable

ML models are integral to a tremendous number of industries, including the automotive, healthcare, financial, and entertainment industries. The need for more accurate and robust ML models drives researchers to propose deeper and more complex ML models. These deep models are usually composed of more layers and thus more neurons. This means a huge number of parameters to tune in the training process. Consequently, ML models need more training data as these industries evolve (see *Figure 5.2*):

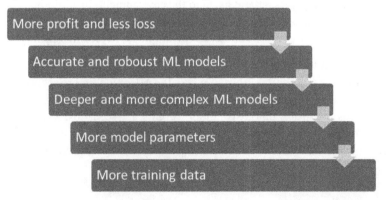

Figure 5.2 – To increase profitability, companies require precise and complex
ML models, which necessitates larger amounts of training data

Unfortunately, the process of collecting, cleaning, and annotating real data is extremely slow and expensive. Companies need to respond to market changes swiftly; otherwise, they may lose customers, reputation, and opportunities.

Synthetic data is perfectly scalable: once the data generation and annotation pipelines have been configured, it is easy to generate large-scale datasets under new conditions as necessary. For example, you can generate an unlimited number of training images using **Generative Adversarial Networks (GANs)**, **Variational Autoencoders (VAEs)**, or simulators such as **CARLA** (`https://carla.org/`).

Automatic data labeling

One of the main advantages of synthetic data is automatic data labeling. Since the data generation process is controlled, automatic data labeling is possible with synthetic data. For example, if you utilize a game engine such as **Unreal** or **Unity** to generate and label synthetic data, it is possible to tag objects and thus to know exactly which objects are seen by your camera at a given frame.

This is the benefit of synthetic data! It saves you extensive time, effort, and money that you need to spend on annotating real data. Moreover, you can annotate private or confidential data without being worried about annotators disclosing sensitive information.

Annotation quality

Synthetic data is automatically annotated, unlike real data. Human factor errors are minimized and limited. Thus, annotating high-quality and large-scale synthetic datasets is possible. For example, synthetic data algorithms can provide, to the pixel level, accurate ground truth for semantic segmentation, which is impossible to achieve with real data due to the limitations of human annotators. In many situations, there is a trade-off between quality and quantity when working with real data. Using synthetic data, you can achieve both objectives with less time and a lower budget.

Low cost

Synthetic data, as we have mentioned, does not require you to capture data from the real world and it does not need annotators to annotate it. Synthetic data can be generated and automatically annotated using the appropriate algorithms. Thus, after developing the generation and annotation pipelines, it is possible to generate an unlimited number of training examples. For example, generating a training dataset of a thousand images and generating a training dataset of a million images would cost almost the same amount!

In the next section, we will look at a particular benefit of synthetic data – that is, using it to solve privacy issues with sensitive data and applications.

Solving privacy issues with synthetic data

In certain fields, such as healthcare and finance, a lot of data is available, but the main obstacle is annotating and sharing the data. Even if we have a large-scale real dataset that is "perfectly" annotated, sometimes, we cannot share it with ML practitioners because it contains sensitive information that could be used by a third party to identify individuals or reveal critical information about businesses and organizations.

As we know, ML models cannot work without data, so what is the solution? A simple solution is to use the *real* data to generate *synthetic* data that we can share with others without any privacy issues while still representing the real data. We can utilize some synthetic data generation approaches to leverage the real dataset to generate a synthetic dataset that still represents the relationship between variables, hidden patterns, and associations in the real data while not revealing sensitive information:

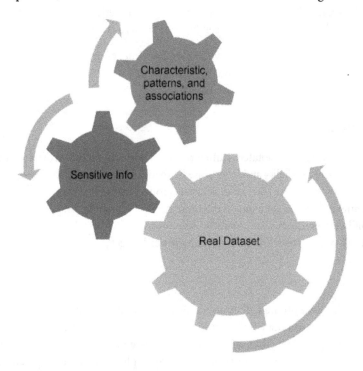

Figure 5.3 – Some synthetic dataset generation approaches
disentangle sensitive information from data patterns

In this scope, we can understand that synthetic data generation approaches disentangle sensitive information from the associations and relationships between the variables (see *Figure 5.3*). Thus, ML models can still be trained on the synthetic data and learn the hidden patterns in the original real data without being directly trained on it.

Synthetic data does not contain information about real individuals, so no harm is caused. For example, let's assume you have a synthetic dataset of human faces. It is fine to use this data as you want. Additionally, you would not be restricted by regulations as you would if you were using real human faces. This would allow you to explore novel creative ideas. However, if you were going to use sensitive information from real humans, you would be limited to the main purpose you used to collect the data for. Thus, you cannot investigate new ideas without permission from the people who participated. Additionally, data should not be kept longer than necessary. All of these regulations limit the usability of sensitive real datasets, which makes synthetic ones a perfect alternative. For more information, please check out *The Data Protection Act 2018*, which is the UK's implementation of the **General Data Protection Regulation (GDPR)** (`https://www.gov.uk/data-protection`).

In the next section, we will discuss why synthetic data is the solution for cost and time-efficiency issues in data generation and annotation.

Using synthetic data to solve time and efficiency issues

Automatic data generation of synthetic data removes many unnecessary elements in the real data curation and annotation pipeline. Collecting real data often requires special equipment, such as high-resolution cameras, microphones, or LiDAR. At the same time, you need engineers and technicians who are trained to use such equipment. You lose time and money training engineers and buying or renting this equipment. Often, data curators need to travel and visit various locations to collect suitable data, meaning that you would have to pay for transportation, accommodation, insurance, and more.

Synthetic data is an effective solution for these issues (see *Figure 5.4*). In addition to the preceding issues, it is easy to conclude that synthetic data has a lower carbon footprint than real data. Thus, it is even better for the environment!

Data annotation is one of the main issues that makes real datasets cumbersome. Annotating large-scale datasets is an extremely expensive and time-consuming process. Annotating a large-scale dataset can take weeks, months, or even years. The huge amount of time it takes to fulfill the annotation process may make companies fall behind their competitors, causing them to lose market share and customers.

It becomes even worse if you are working with real sensitive data. By law, it is mandatory to take extra care when it comes to storing, processing, annotating, or transferring this type of data. This means more budget, effort, and time to spend. However, using synthetic data removes all of this extra work and eases your workload:

	Real Data	Synthetic Data
Recording/Capturing Equipment	↑	↓
Transportation/Accommodation	↑	↓
Training	↑	↓
Insurance	↑	↓
Time	↑	↓
Regulations	↑	↓
↑ You need more and ↓ you need less		

Figure 5.4 – Comparison between synthetic and real data

In the next section, we will understand why synthetic data can cover rare and special scenarios compared to real data.

Synthetic data as a revolutionary solution for rare data

Rare data occurs in the real world because of infrequent events or phenomena. In other words, these events occur but with low frequency. We can broadly classify these events into these categories:

- **Natural catastrophes**: This category includes events such as floods, asteroid impacts, earthquakes, and tsunamis

- **Anthropogenic**: This category includes events such as industrial accidents, financial crises, and violent conflicts

These events create many major changes in the environment, which may cause state-of-the-art ML models to fail. For example, a face recognition system may not work well in the case of the evacuation of a building because as the building becomes more crowded, movement patterns may change. While these events are rare, their impacts on societies are tremendous. ML models that function inappropriately may greatly increase the number of deaths and injuries.

For ML models to be robust and accurate, these models need to be trained and tested on both standard and rare conditions. Capturing real data of rare events is extremely hard and expensive. Most ML models assume that they will work under standard conditions and scenarios. Unfortunately, these ML models generally fail or struggle under any scenarios that deviate from the standard ones. For instance, in *Semantic Segmentation under Adverse Conditions: A Weather and Nighttime-aware Synthetic Data-based Approach* (https://bmvc2022.mpi-inf.mpg.de/0977.pdf), researchers demonstrated that state-of-the-art semantic segmentation methods perform well under standard conditions, such as normal weather conditions and sufficient illumination. However, these methods struggle or fail under adverse conditions, such as foggy, rainy, and snowy weather conditions or at nighttime.

The key reason for neglecting rare scenarios is the fact that collecting training data under these circumstances may take a long time, a lot of training and effort is required to capture these rare events, and it can be a dangerous process. Finally, we should note that rare data is not just useful for training purposes – it is essential for understanding the limitations of ML models in practice.

As mentioned previously, real data is not balanced in the real world. Thus, the data that we collect from the real world will reflect this imbalance. Unfortunately, ML models are sensitive to imbalanced datasets. Thus, imbalanced training datasets cause ML models to develop a corrupted understanding of the problem. For example, if we train a cats-dogs classifier on a dataset that includes 30 cats and 70 dogs, the model will tend to predict dogs twice as often as it predicts cats. Thus, balanced training datasets make the models train better and converge faster.

Standard conditions, events, and attributes are more likely to occur in the real world than rare events. Thus, you are more likely to have a dataset that specifically focuses on normal conditions and neglects rare ones.

As you might expect, synthetic data can be used to simulate these rare events. Thus, generating a perfectly balanced large-scale dataset is easy to achieve with synthetic data. Synthetic data generation methods may be used to generate a full training dataset from scratch. At the same time, synthetic data can be utilized to complement real datasets. Thus, you can make your ML models more robust against rare events and conditions.

Synthetic data generation methods

There are different methods to generate synthetic data: some of them are based on statistical models and others rely on game engines and simulators. *Statistical models* are non-deterministic mathematical models that include variables represented as probability distributions. Based on the problem, these models are usually trained using real data to understand the hidden patterns and correlations in the data. Then, the trained ML model can be used to generate new samples automatically, such as images, text, tables, and more. These new samples can be utilized by other ML models for training or testing purposes.

Synthetic data can also be generated using *game engines and simulators*. These tools are utilized to create 3D virtual worlds. These 3D worlds can be generated using **Procedural Content Generation** (**PCG**) techniques to control scene attributes, the interaction between scene elements, and the diversity and quality of the generated data.

It is important to note that the main challenge for most synthetic data generation approaches is building a data generation and annotation pipeline, which requires careful design and engineering. However, once the pipeline is ready, it is usually simple to use and can be utilized for an enormous range of applications.

Summary

In this chapter, we learned about the main advantages of using synthetic data. We discussed that synthetic data is easy to generate, manage, and annotate. When it comes to privacy issues that we have with sensitive real data, utilizing synthetic data is an ideal solution.

In the next chapter, we will learn how to utilize simulators and rendering engines to generate synthetic data.

Part 3:
Synthetic Data Generation Approaches

In this part, you will be introduced to the main synthetic data generation approaches. You will learn how to leverage simulators and rendering engines, **Generative Adversarial Networks** (**GANs**), video games, and diffusion models to generate synthetic data. You will explore the potential of these approaches in ML. Moreover, you will understand the challenges and pros and cons of each method. This part will be supported with hands-on practical examples to learn how to generate and utilize synthetic data in practice.

This part has the following chapters:

- *Chapter 6, Leveraging Simulators and Rendering Engines to Generate Synthetic Data*

- *Chapter 7, Exploring Generative Adversarial Networks*

- *Chapter 8, Video Games as a Source of Synthetic Data*

- *Chapter 9, Exploring Diffusion Models for Synthetic Data*

6

Leveraging Simulators and Rendering Engines to Generate Synthetic Data

In this chapter, we will introduce a well-known method for synthetic data generation using simulators and rendering engines. We will explore the main pipeline for creating a simulator and generating automatically annotated synthetic data. Following this, we will highlight the challenges and briefly discuss two simulators for synthetic data generation.

In this chapter, we're going to cover the following main topics:

- Simulators and rendering engines: definitions, history, and evolution
- How to generate synthetic data
- Challenges and limitations
- Case studies

Introduction to simulators and rendering engines

In this section, we will dive into the world of simulators and rendering engines. We will look at the history and evolution of these powerful tools for synthetic data generation.

Simulators

A **simulator** is software or a program written to imitate or simulate certain processes or phenomena of the real world. Simulators usually create a virtual world where scientists, engineers, and other users can test their algorithms, products, and hypotheses. At the same time, you can use this virtual environment to help you learn about and practice complex tasks. These tasks are usually dangerous and very expensive to perform in the real world. For example, driving simulators teach learners how to drive and how to react to unexpected scenarios such as a child suddenly crossing the street, which is extremely dangerous to do in the real world.

Simulators are used in various fields, such as aviation, healthcare, engineering, driving, space, farming, and gaming. In *Figure 6.1*, you can find examples of these simulators.

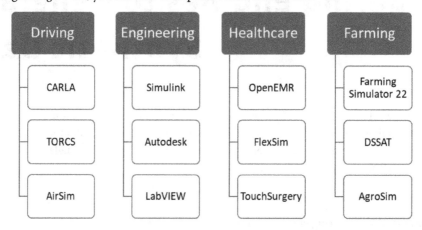

Figure 6.1 – Examples of simulators utilized in driving, engineering, healthcare, and farming

Next, we will introduce rendering and game engines.

Rendering and game engines

Rendering and game engines are software used mainly to generate images or videos. They are composed of various subsystems responsible for simulating, for example, physics, lighting, and sound. They are usually used in fields such as gaming, animation, virtual reality, augmented reality, and the metaverse. Unlike simulators, game engines can be used to create virtual worlds that may or may not be designed to mimic the real world. Game engines are mainly used to develop games. However, they can be utilized for training and simulation, films and television, and visualization. In *Figure 6.2*, you can see some examples of modern rendering and game engines.

Figure 6.2 – Examples of modern rendering and game engines

Next, we'll learn more about the history of rendering and game engines.

History and evolution of simulators and game engines

Game engines roughly started to appear in the 1970s. The computers of that era were limited in terms of processing capabilities and memory. At that time, most games were 2D games, such as *Pong* and *Spacewar!*. They were limited to simple graphics, basic lighting, elementary shading, and limited visual and sound effects.

The great advancements in hardware led to more sophisticated game engines. These game engines, such as *Unreal* (https://www.unrealengine.com) and *Unity* (https://unity.com) facilitated the creation of rich, photorealistic virtual worlds. As physics simulations became more sophisticated and advanced, the development of photorealistic graphics also progressed simultaneously. This allowed the simulation of complex physics and interactions between scene elements, such as fluid dynamics and cloth simulation. In more recent years, many photorealistic, complex games have been released, such as *Call of Duty* (https://www.callofduty.com) and *Grand Theft Auto* (https://www.rockstargames.com/games/vicecity).

The availability of complex and easy-to-use game engines such as Unity has democratized game development and made it more accessible than ever before. Thus, game development is not just limited to big tech companies but is also available to independent companies and artists.

At the same time, the huge sudden increase in the number of mobile phones recently made mobile games a more attractive destination for research and industry.

Recently, synthetic data researchers started to experiment with using game engines and video games to generate rich synthetic data. Two of the pioneer works in this area are *Playing for Data: Ground Truth from Computer Games* (`https://arxiv.org/abs/1608.02192`) and *Domain Randomization for Transferring Deep Neural Networks from Simulation to the Real World* (`https://arxiv.org/pdf/1703.06907.pdf`).

In the following section, we will explore exactly how to generate synthetic data using simulators and game engines.

Generating synthetic data

In this section, we will learn how to generate synthetic data using modern game engines such as Unity and Unreal.

To generate synthetic data with its corresponding ground truth, it is recommended that we follow these steps:

1. Identify the task and ground truth to generate.
2. Create the 3D virtual world in the game engine.
3. Set the virtual camera.
4. Add noise and anomalies.
5. Set the labeling pipeline.
6. Generate the training data with the ground truth.

Throughout this section, we will thoroughly discuss each facet of this process.

Identify the task and ground truth to generate

The first step in the synthetic data generation process is defining the task, the type of the data, and the ground truth to generate. For example, the data could be images, videos, or audio. At the same time, you need to identify what ground truth to generate for your problem. For example, you can generate semantic segmentation, instance segmentation, depth maps, normal maps, human poses, and human body parts semantic segmentation, just to name a few.

Next, we need to understand how to create the 3D virtual world, which we will explore in the following section.

Create the 3D virtual world in the game engine

To begin with, we must define the environment, its elements, and the interactions between these elements. You may need to decide on the level of photorealism, the degree of visual complexity, and the range of variations and diversity that you need to attain for your virtual scenes, and thus the synthetic data. In general, and for a typical plan to generate synthetic data, we can follow these steps:

1. Preparation and conceptualization

2. Modeling

3. Materialization and texturing

4. Integration into the game engine

5. Polishing and testing

Next, we will delve into each of these aspects and provide deeper insight.

Preparation and conceptualization

Before creating the 3D virtual world, we need to examine our ideas about the virtual world to be created. It is suggested to make simple drawings and sketches to visualize the elements of the world and how they will interact with each other. You may need to jot down the following: weather conditions to simulate, whether an indoor or outdoor environment, the time of day, and the scenes' crowdedness, just to mention a few. Additionally, you need to decide which game engine to use, for instance, **Unity**, **Unreal**, or **CryEngine**. You also need to decide which rendering pipeline to utilize, which depends on the game engine itself. For example, the Unity game engine has different rendering pipelines, such as **Built-in Render Pipeline (BRP)**, **High-Definition Render Pipeline (HDRP)**, **Universal Render Pipeline (URP)**, and **Scriptable Render Pipeline (SRP)**. The selection of the rendering pipeline also depends on the degree of photorealism that you want to achieve. Moreover, some game engines may support various programming languages, such as CryEngine, which supports C++ and C#. Thus, you may need to decide which language to use as well.

After this, we need to determine the assets to use, such as objects, materials, visual effects, and sound effects. At the same time, you may need to consider the budget, timeframe, and the skills of your team.

Modeling

The next step, after creating a solid idea about the 3D virtual world, is to start the modeling stage. **3D modeling** is the process of creating 3D objects using appropriate modeling software. 3D modeling is widely used in the game and entertainment industries, engineering fields, and architecture. To build the 3D virtual world, we need to create its elements, such as buildings, trees, pedestrians, and vehicles. Thus, we need to decide whether to import or model these elements. We can do 3D modeling using software such as *Blender* (https://www.blender.org), *ZBrush* (https://pixologic.com), and *3ds Max* (https://www.autodesk.co.uk/products/3ds-max). As you may expect, a straightforward solution is importing these elements from websites such as *Adobe 3D Substance* (`https://substance3d.adobe.com`) and *Turbosquid* (`https://www.turbosquid.com`). However, high-quality 3D models are usually expensive. Additionally, it should be noted that modeling complex 3D objects can be a challenging and time-consuming process that requires technical skills, effort, and time, but that depends on the object being modeled and technical constraints such as the polygon count.

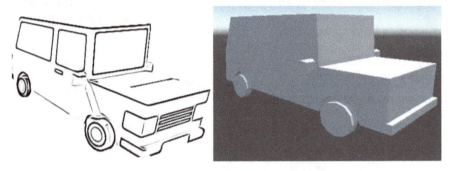

Figure 6.3 – An example of a 3D car model (right) created from a car sketch (left)

Figure 6.3 shows an example of the output that we get after the modeling stage, which is a car in this instance.

Materialization and texturing

After creating a 3D model or a mesh, we need to add the physical properties of this object, such as the color, transparency, and reflectivity. These properties simulate the matter and the surface of the objects. On the other hand, texturing is used to simulate surface details such as scratches and patterns and to give the object a non-uniform appearance similar to the object's appearance in the real world. In most game engines, this information is encoded using a texture map. *Figure 6.4* shows a 3D object after the materialization and texturing process.

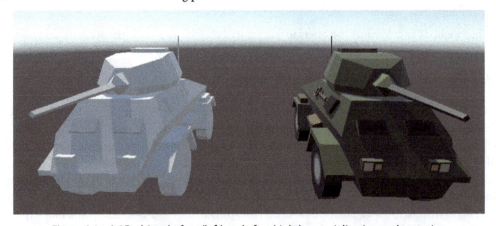

Figure 6.4 – A 3D object before (left) and after (right) materialization and texturing

Integration into the game engine

When the elements of the 3D virtual world are ready, we need to add them to our scene. We also need to configure and set up lighting and a virtual camera:

- **Lighting**: This is an essential step for creating photorealistic scenes. Lights are added to the virtual worlds to give a sense of depth and atmosphere. There are usually two options for lighting: **pre-rendered lighting** using lightmaps and **real-time lighting**. Lighting is fundamental but it is expensive computation-wise. Thus, you should pay attention to this step to achieve your target photorealism and frame rate.

- **Virtual camera**: Once the virtual world is generated, a virtual camera is utilized to capture the required synthetic data. The behavior of the camera is usually controlled using a script. The camera parameters and behaviors can be configured to match the real-world scenario and to achieve the intended behavior. Camera parameters include **Field of View (FoV)**, **Depth of Field (DoF)**, **Sensor Size**, and **Lens Shift**.

Polishing and testing

The last step is to examine the generated synthetic data and iterate on the virtual world design. In this stage, you can fix bugs and optimize the performance. As expected, creating a 3D virtual world is not a simple process. It requires effort, time, and technical skills. However, once the virtual world is created, it can be leveraged to generate large-scale synthetic datasets for enormous applications.

Setting up the virtual camera

In the virtual world, the camera plays a vital role in the synthetic data generation process. It represents the observer, and it is usually utilized to capture images, audio, and videos. The captured data may be used for training and testing ML models.

As we have mentioned previously, camera properties and attributes can be customized and configured to achieve the target behavior. For example, the camera FoV controls how much of the world your observer agent can perceive, and therefore, how much visual information you can capture in a single generated image. *Figure 6.5* shows two images generated with different FoV values.

Figure 6.5 – A scene captured using two different FoVs in the Unreal game engine

Please note that the camera position is fixed, and we only changed the FoV. Additionally, we can control the camera motion and transition to achieve the required behavior.

The camera can take different setups in the 3D virtual world to imitate the relevant ones in the real world. These are some examples of camera setups:

- **Fixed camera**: The camera does not change its location or orientation while it captures the scene. This camera setup can be used to record the scene from a specific viewpoint. It is the simplest setup; it is easy to implement, and it does not require scripting. However, you need to pay attention to the position and the attributes of the camera. Otherwise, dynamic objects may accidentally block the view of the camera. In simulators and game engines, a fixed camera can be used, for instance, to imitate a traffic monitoring camera or a fixed camera used in sports broadcasting.

- **PTZ camera**: This is a special type of camera setup in which the camera can pan, tilt, and zoom. In the real world, this type is usually controlled by an operator to capture an object of interest or a specific event. Thus, the camera can change its orientation and FoV to realize that. In a virtual world, the camera can be programmed to achieve that, or it can be controlled by a human operator during simulation or the synthetic data generation process. This setup gives you more freedom to capture the scene. However, it may require scripting to achieve the intended camera behavior.

- **First-person camera**: First-person vision is a fundamental field in computer vision. This camera setup simulates an agent observing the world by wearing a camera. It has enormous applications in gaming and virtual reality, law enforcement, medicine, and education. For example, an ML model trained on first-person data can be used to assist surgeons and improve training, decision-making, performance, and accuracy. For a detailed discussion, refer to *Artificial Intelligence for Intraoperative Guidance: Using Semantic Segmentation to Identify Surgical Anatomy During Laparoscopic Cholecystectomy* (https://pubmed.ncbi.nlm.nih.gov/33196488).

- **Aerial or UAV camera**: This camera setup is key for flight and drone simulators. It simulates a camera mounted on a drone or UAV. Usually, it is used to simulate a birds-eye view of the scene. It has a wide spectrum of applications and can be used to enhance the performance of ML models that require training images captured using drones. It supports various computer vision tasks, such as object detection, classification, tracking, and recognition.

- **Stereoscopic camera**: A stereo camera is a special type of camera with two lenses separated by a short distance that can be leveraged to simulate how humans perceive depth. The distance between the two lenses is called **intra-ocular distance** and it is usually similar to the distance between a human's eyes: approximately 6.35 cm. This distance is essential for creating a sense of depth in the vision system. This type of camera is important for VR and immersive 3D experiences.

- **Tracking camera**: This type is used to track an object of interest. In the virtual world, this camera can be programmed to follow the desired object, which facilitates creating large-scale synthetic data focused on a target object. For example, it is possible to track a human in the virtual world for an action recognition task. This will help you to generate large-scale training data focused on your subject (human). It is possible to use other camera setups, but you will end up having many videos with no actions or where your object of interest is not apparent. Additionally, you may use this camera setup for visual object tracking and other similar tasks.

The next step is adding noise.

Adding noise and anomalies

The real world is not perfect and has anomalies. In the context of image generation, noise and anomalies refer to a deviation from the main pattern, process, and phenomenon. For example, when we observe street light poles at night, a small portion of them may have been accidentally turned off, the light may be flickering, the pole may be slightly rotated, painted a different color, or have different dimensions. Adding noise and anomalies to the attributes and behaviors of virtual world elements improves realism and boosts the usability of the generated synthetic data. Another example regarding anomalies in behavior can be seen when observing pedestrians crossing the road. The majority wait for the green light or walk signal, look both ways, and cross on the crosswalk. On the other hand, a small portion may cross the road when the red light is on or may cross without paying attention to oncoming cars. This behavior anomaly, for example, should be simulated in the virtual world to ensure that the generated training data is diverse. Thus, training an ML model on this data will ensure a robust ML model.

Setting up the labeling pipeline

The labeling pipeline depends on your problem and how you plan to utilize the synthetic data. You may want to just generate the training data because it is too expensive in the real world, and you may prefer to ask human annotators to annotate your data. On the other hand, it is possible that you want to automate the annotation process in the simulator or the rendering engine. Simulators such as CARLA, NOVA, and Silver support generating the data with its corresponding ground truth for various computer vision tasks.

Generating the training data with the ground truth

At this point, the synthetic data generation pipeline should be ready. The previous steps can be challenging, costly, and time-consuming. However, they are only required to set up the system. Following this, you can leverage the system to generate your task-specific, automatically annotated, large-scale datasets. Changing the annotating protocol is simple and is not expensive compared to real-world datasets. Please note that we have not provided hands-on examples on how to generate synthetic data using game engines or simulators because the focus of the book is not on the implementation and coding of synthetic data generation approaches. However, it is committed to the theoretical, conceptual, and design aspects of the process. For more details about the implementation and coding aspects, please refer to Unity Computer Vision (`https://unity.com/products/computer-vision`) and Synthetic for Computer Vision (`https://github.com/unrealcv/synthetic-computer-vision`).

In the next section, you will learn about the main limitations of deploying this synthetic data generation approach.

Challenges and limitations

In this section, we will highlight the main challenges in using this approach for synthetic data generation. We will look at realism, diversity, and complexity issues that present some difficulties in utilizing this approach for synthetic data generation.

Realism

The **domain gap** between synthetic and real data is one of the main issues that limit the usability of synthetic data. For synthetic data to be useful, it should mimic the distribution and statistical characteristics of its real counterparts. Thus, for computer vision problems, we need to ensure a high degree of photorealism, otherwise, ML models trained on synthetic data may not generalize well to real data.

Achieving a high degree of photorealism using game engines and simulators is not a simple task. Even with the help of contemporary game engines such as CryEngine, Unreal, and Unity, we need effort, skill, and time to create photorealistic scenes.

The three essential elements for approaching realism and thus mitigating the domain gap problem for synthetic data generated by game engines and simulators are as follows.

Photorealism

Generating photorealistic images, for computer vision problems, is vital for training and testing ML models. Building photorealistic scenes requires the following:

- **High-quality assets**: The 3D models, textures, and materials should be detailed and realistic.
- **Lighting**: It is essential for rendering photorealistic scenes. You may need to use physically based rendering and physically based materials. Additionally, you need to use suitable light sources and carefully configure their parameters.
- **Postprocessing**: Game engines such as Unreal and Unity support postprocessing effects to improve photorealism. For example, you can use these techniques to simulate motion blur, gloom, and depth of field.

Realistic behavior

To achieve this, we need to ensure realistic camera behavior. For example, the camera should not penetrate walls and its parameters should be close to real-world camera ones. Additionally, character animations should be realistic and emulate human body movement. Furthermore, scene element interactions should obey physics rules, for example, when objects collide.

Realistic distributions

Objects and attributes are not randomly distributed in the real world. Thus, when building virtual worlds, we need to pay attention to matching these distributions as well. For example, people walking near shopping malls may have a higher probability of carrying objects. At the same time, under certain weather conditions, specific actions and objects may become more frequent or less. For example, in rainy weather conditions, pedestrians may carry umbrellas.

Diversity

The world around us is remarkably varied and contains a multitude of elements that come in diverse colors, shapes, and behaviors, and possess different properties. Attaining a diverse virtual world requires time and effort. The usability of synthetic data comes from its primary advantage of generating large-scale datasets for training ML models. If the data is not diverse enough, this will cause ML models to overfit limited scenarios and attributes.

Complexity

Nonlinearity, interdependence, uncertainty, and the dynamic nature of the real world make creating a realistic virtual world rather a complex task. Creating and simulating a realistic environment requires approximations, simplifications, and generalizations. This is necessary because of the trade-off between realism and computational complexity. Building a realistic virtual world that captures all real-world properties, phenomena, and processes is simply not feasible, even with state-of-the-art software and hardware. However, we can still approach an acceptable level of realism with a careful understanding of the ML problem: what is essential and what is auxiliary for this particular application? For example, if we would like to generate synthetic data for a face recognition task, we may need to pay extra attention to simulating photorealistic faces as compared to other elements in the scene.

Looking at two case studies

In this section, we will briefly discuss two well-known simulators for synthetic data generation, and comment on the potential of using these approaches.

AirSim

AirSim is an open source, cross-platform simulator developed by Microsoft using the Unreal game engine. It simulates drones and cars, opening the door for enormous applications in computer vision for DL and RL approaches for autonomous driving. Some of the key features of this simulator include the following:

- Various weather effects and conditions

- LIDAR and infrared sensors

- Customizable environment

- Realistic physics, environments, and sensors

As you can see, AirSim can be leveraged to generate rich, large-scale, and high-quality synthetic data from various sensors. Researchers in ML can train their models to fuse the different data modalities to develop more robust autonomous driving algorithms. Additionally, AirSim provides automatically labeled synthetic data for depth estimation, semantic segmentation, and surface normal estimation tasks. For more information about this simulator, please refer to *AirSim* (`https://www.microsoft.com/en-us/AI/autonomous-systems-project-airsim`).

CARLA

CARLA is an open source simulator for autonomous driving. It was developed using the Unreal game engine by **Computer Vision Centre (CVC)**, Intel, and Toyota. CARLA is a well-known simulator for synthetic data generation. It has a traffic manager system and users can configure several sensors, which include depth sensors, LIDARs, multiple cameras, and **Global Positioning System (GPS)**. CARLA generates synthetic data for a number of computer vision tasks, such as semantic segmentation, depth estimation, object detection, and visual object tracking. In addition to generating automatically labeled and large-scale synthetic data, CARLA can be deployed to generate diverse traffic scenarios. Then, researchers can utilize the generated synthetic data to train more accurate and robust ML models on a myriad of driving scenarios. Please check the project's *CARLA* web page (`https://carla.org`) and the *CARLA: An Open Urban Driving Simulator* paper (`http://proceedings.mlr.press/v78/dosovitskiy17a/dosovitskiy17a.pdf`) for more details.

There are many other simulators, such as Silver, AI Habitat, SynthCity, and IGibson. Creating a more realistic simulator, supporting more tasks, making the simulator easier to use, and the virtual environment more customizable are the main research directions in developing future synthetic data generators using game engines and simulators.

Summary

In this chapter, we introduced a well-known method for synthetic data generation based on simulators and rendering engines. We learned how to generate synthetic data. We highlighted the main challenges and we discussed AirSim and CARLA simulators as examples of this data generation approach. We have seen that by using simulators and game engines, we can generate large-scale, rich, and automatically annotated synthetic data for many applications. It reduces the cost and effort and provides an ideal solution for training robust ML models.

In the next chapter, we will learn about a new method for synthetic data generation using **Generative Adversarial Networks (GANs)**.

7

Exploring Generative Adversarial Networks

In this chapter, we will introduce **Generative Adversarial Networks** (**GANs**) and discuss the evolution of this data generation method. You will learn about the typical architecture of a GAN. After this, we will explain its training process and discuss the main challenges. Then, we will highlight various applications of GANs, including generating images and text-to-image translation. Additionally, we will study a practical coding example demonstrating how to use GANs to generate photorealistic images. Finally, we will also discuss variations of GANs, such as conditional GANs, CycleGANs, CTGANs, WGANs, WGAN-GPs, and f-GANs.

In this chapter, we're going to cover the following main topics:

- What is a GAN?
- Training a GAN
- Utilizing GANs to generate synthetic data
- Hands-on GANs in practice
- Variations of GANs

Technical requirements

The code used in this chapter will be available in the corresponding chapter folder in the book's GitHub repository: `https://github.com/PacktPublishing/Synthetic-Data-for-Machine-Learning`.

What is a GAN?

In this section, we will introduce GANs and briefly discuss the evolution and progression of this particular data generation method. Then, we will explain the standard architecture of a typical GAN and how they work.

The concept of GANs was introduced in the 2014 paper *Generative Adversarial Networks* (`https://arxiv.org/abs/1406.2661`), by Ian J. Goodfellow and his research team. In the same year, **conditional GANs** were introduced, allowing us to generate more customizable synthetic data. Then, **Deep Convolutional GANs (DCGANs)** were suggested in 2015, which facilitated the generation of high-resolution images. After that, **CycleGANs** were proposed in 2017 for unsupervised image-to-image translation tasks. This opened the door for enormous applications such as domain adaptation. **StyleGAN** was introduced in 2019, bringing GANs to new fields such as art and fashion.

GANs have also been showing impressive progress in the field of video synthesis. In fact, the recent work by NVIDIA is a testament to their tremendous potential (please check this paper for more details: *One-Shot Free-View Neural Talking-Head Synthesis for Video Conferencing* at `https://arxiv.org/pdf/2011.15126.pdf`).

This work shows that GANs can now recreate a talking-head video using only a single source image. For the code, dataset, and online demo, refer to the project's page: `https://nvlabs.github.io/face-vid2vid`. Next, we delve into the architecture of GANs.

Most **deep learning (DL)** methods and architectures are designed to predict something. It could be weather conditions, stock prices, object classes, or something else. However, GANs were proposed to *generate* something. It could be images, videos, texts, music, or point clouds.

At the heart of this capability lies the essential problem of learning how to generate training samples from a given domain or dataset. GANs are DL methods that can learn complex data distributions and can be leveraged to generate an unlimited number of samples that belong to a specific distribution. These generated synthetic samples have many applications for data augmentation, style transfer, and data privacy.

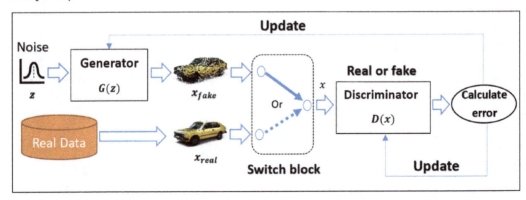

Figure 7.1 – A typical architecture and training process of GANs

Moving forward, we will learn how to train GANs.

Training a GAN

In this section, we will learn how to train a typical GAN. Then, we will discuss the main challenges and difficulties.

A GAN is trained using **unsupervised learning** techniques where both submodels are trained simultaneously using a process called **adversarial training**. A typical GAN consists of two neural networks (usually convolutional neural networks): the **generator** and the **discriminator**. The generator takes in a random noise vector as input and generates a synthetic (fake) sample. The goal of the generator is to produce synthetic data that is realistic and indistinguishable from real data. The discriminator, on the other hand, is trained to distinguish between real and fake samples. It receives a sample and predicts its data source domain: real or fake. If the discriminator correctly identifies a real data sample, no error is backpropagated. However, if the discriminator fails to identify a synthetic sample, it is penalized, and the generator is rewarded. The generator is penalized if the discriminator is able to correctly identify generated, synthetic data. In this way, both the generator and discriminator are constantly trying to improve their performance, resulting in the generation of increasingly realistic synthetic data. Refer to *Figure 7.1* for a visualization of the training process. Let's explore more and learn about the training process in more detail.

> **Disclaimer on hands-on training of GANs**
>
> Please note that we do not provide hands-on elements on how to train GANs because the chapter is committed to the theoretical, conceptual, and design aspects of GANs for synthetic data generation. Thus, hands-on examples are out of the scope of this chapter. However, if you are keen to train your GAN, please refer to the *Deep Convolutional Generative Adversarial Network Tutorial* (https://www.tensorflow.org/tutorials/generative/dcgan).

GAN training algorithm

The training algorithm is a crucial aspect of enabling GANs to generate useful synthetic data. The following is a step-by-step procedure that can be utilized to train GANs:

1. Create z by sampling a random noise following a suitable noise distribution such as uniform, Gaussian, Binomial, Poisson, Exponential, Gamma, and Weibull distributions.

2. Feed z to the generator to produce a synthetic or fake sample, *x fake*.

3. Pass both *x fake* and *x real* to a `switch` block, which randomly selects one of its inputs and passes it to the discriminator.

4. The discriminator classifies the given sample as real or fake.

5. Calculate the error.

6. Backpropagate the error to both the generator and discriminator.

7. Update the weights of the generator and discriminator

Next, we'll discuss the loss.

Training loss

The loss shown next is one of many losses that can be used to train a GAN. This particular loss is derived from the **cross-entropy loss**:

$$L = E\,x[\log(D(x))] + E\,z[\log(1 - D(G(z)))]$$

Let's break this formula down:

- $D(x)$ is the discriminator's estimate that x is drawn from the real dataset

- $E\,x$ and $E\,z$ are the expected values over real and generated synthetic (fake) samples, respectively

- $G(z)$ is the output of the generator for a noise vector, z

- $D(G(z))$ is the discriminator's estimate that a synthetic sample is real

As expected, the training process of GANs is complex, but it is a powerful technique for generating realistic data, which motivated researchers to examine new ways to enhance and speed its training and convergence. Next, let us discuss some of these challenges.

> **GANs in action**
>
> For an interactive demonstration of how a GAN is trained, please refer to *Play with Generative Adversarial Networks (GANs)* in your browser (`https://poloclub.github.io/ganlab`). For more details, check out the corresponding paper *GAN lab: Understanding Complex Deep Generative Models using Interactive Visual Experimentation* (`https://minsuk.com/research/papers/kahng-ganlab-vast2018.pdf`).

Challenges

Now we will cover some common issues and challenges encountered when training a GAN. Let's explore insights and explanations about the cause of such issues:

- **Mode collapse:** In this scenario, the generator overfits to a limited number of samples and patterns producing the same or similar synthetic samples for different z values. For example, a GAN being trained to generate cat images may keep generating the same cat image again and again with just minor modifications. This is something that we do not want to happen. The point of using GANs is to generate diverse synthetic examples. This problem occurs when the generator learns to produce one or a few excellent synthetic samples that fool the discriminator. Thus, the generator avoids generating other samples and prefers to repeat these excellent synthetic samples. There are various solutions to this problem, such as *unrolled GANs* (`https://arxiv.org/pdf/1611.02163.pdf`) and *Wasserstein loss* (`https://arxiv.org/abs/1701.07875`).

- **Discriminator saturations (diminished gradients):** As we discussed earlier, the training of the generator and discriminator is done in an adversarial manner. When the discriminator becomes too successful at classifying real from synthetic samples, the error becomes minimal. Thus, the generator can no longer learn useful things.

- **Hyperparameter sensitivity and tuning:** Similar to other DL architectures, GANs have many hyperparameters, such as learning rate, batch size, number of layers, activation functions, and others. Finding the optimal hyperparameters is problem- and task-dependent and usually a train-error process. Thus, it is challenging to find the right architecture and hyperparameters to successfully train your GAN.

- **Instability and non-convergence:** It is not easy to stabilize the training process of the generator and discriminator. In fact, it is common to observe that one submodel is learning better than another, which causes the GAN to oscillate, giving us unpredictable behavior, and the models may never converge. For more details, please refer to *On Convergence and Stability of GANs* (`https://arxiv.org/pdf/1705.07215.pdf`).

- **Computation complexity:** GANs have a complex structure, being composed of two DL models. This makes the training process computationally expensive and time-consuming. However, there are some techniques proposed to speed up the training process, such as *Small-GAN: Speeding up GAN Training using Core-Sets* (`http://proceedings.mlr.press/v119/sinha20b/sinha20b.pdf`) and *Projected GANs Converge Faster* (`https://proceedings.neurips.cc/paper/2021/file/9219adc5c42107c4911e249155320648-Paper.pdf`).

In the next section, we delve into deploying GANs to generate synthetic data.

Utilizing GANs to generate synthetic data

In this section, we will highlight some interesting applications of GANs.

GANs have enormous applications because they can be used for data augmentation, style transfer, privacy protection, and generating photo-realistic images.

Let's discuss some of these applications:

- **Generating images:** GANs can be utilized to generate photorealistic images. For instance, GANs were utilized to generate handwritten digits, human faces, animals, objects, and scenes. Please check this paper for more details: *Progressive Growing of GANs for Improved Quality, Stability, and Variation* (`https://arxiv.org/pdf/1710.10196.pdf`).

- **Generating cartoon and anime characters:** GANs can be trained to generate appealing and diverse characters. This can be utilized to assess artists, game developers, and designers with anime characters. For more details, please check the paper *Towards the Automatic Anime Characters Creation with Generative Adversarial Networks* (`https://arxiv.org/pdf/1708.05509.pdf`) and the website (`https://make.girls.moe`).

- **Image-to-image translation**: GANs can be utilized to transform images from one domain to another domain. For example, **machine learning** (**ML**)-based colorizers usually utilize GANs for turning grayscale images into colored ones. *Image-to-Image Translation with Conditional Adversarial Networks* (`https://arxiv.org/abs/1611.07004`) and *StarGAN: Unified Generative Adversarial Networks for Multi-Domain Image-to-Image Translation* (`https://arxiv.org/pdf/1711.09020.pdf`) are well-known examples of image-to-image GAN-based translators.

- **Text-to-image translation**: Another interesting application of GANs is to generate appealing images from a given short textual description of scenes and objects. As examples, check *StackGAN: Text to Photo-realistic Image Synthesis with Stacked Generative Adversarial Networks* (`https://arxiv.org/abs/1612.03242`) and DALL-E (`https://openai.com/research/dall-e`).

In addition to the applications we have discussed, GANs can be used for the following non-exhaustive list of interesting tasks and applications:

- Semantic image-to-photo translation

- Generate photographs of human faces

- Face aging

- Pose guided person image generation

- Photos to emojis

- Photograph editing

- Image blending

- Image inpainting

- Super-resolution

- Video prediction

- 3D object generation

- Texture synthesis

- Anomaly detection

Next, we will delve into a hands-on example that demonstrates the practical application of GANs for generating photorealistic synthetic images.

Hands-on GANs in practice

Let's examine how we can utilize a GAN to generate some synthetic images in practice. We will examine *Closed-Form Factorization of Latent Semantics in GANs* (https://arxiv.org/abs/2007.06600) to learn how we can simply generate synthetic images for our ML problem. The code for this example was adapted from the paper's original GitHub (https://github.com/genforce/sefa).

We begin by importing the essential libraries as shown:

```
# import the required libraries
import cv2
import torch
import numpy as np
from utils import to_tensor
from utils import postprocess
from utils import load_generator
from models import parse_gan_type
from utils import factorize_weight
from matplotlib import pyplot as plt
```

Then, we select the parameters of the generation process such as the number of images to generate, and the noise seed. Please note that the seed parameter will help us to get diverse images in this example:

```
num_samples = 1 # num of image to generate (min:1, max:8)
noise_seed = 186 # noise seed (min:0, max:1000)
```

Next, we have the latent semantics parameters of the GAN as proposed by **SeFa**. Simply, we can change some semantics of the synthesized image by changing these parameters. For example, we can change the painting style, gender, posture, and other semantics of the generated image. For more details about **SeFa**, please refer to *Closed-Form Factorization of Latent Semantics in GANs* (https://arxiv.org/abs/2007.06600):

```
# params of generation
layer_idx = "0-1"  # ['all', '0-1', '2-5', '6-13']
semantic_1 = -1.4  # min:-3.0, max:3.0
semantic_2 = -2.9  # min:-3.0, max:3.0
semantic_3 = -1.2  # min:-3.0, max:3.0
semantic_4 = 0.2   # min:-3.0, max:3.0
semantic_5 = -1.4  # min:-3.0, max:3.0
```

Now, we have the following models:

- `stylegan_animeface512`: This can be used to generate anime faces with diverse expressions. For more details, please refer to *A Style-Based Generator Architecture for Generative Adversarial Networks* (`https://arxiv.org/abs/1812.04948`).

- `stylegan_car512`: This can be utilized to generate interesting car models. We will use this model in our example.

- `stylegan_cat256`: We can leverage this model to generate photorealistic cat images.

- `pggan_celebahq1024`: This is a **Progressive Growing GAN** (**PGGAN**) that was trained to generate photorealistic celebrity images. For more details, please refer to *Progressive Growing of GANs for Improved Quality, Stability, and Variation* (`https://arxiv.org/abs/1710.10196`).

- `stylegan_bedroom256`: This can be deployed to generate bedroom layout images. For more details, please refer to *Analyzing and Improving the Image Quality of StyleGAN* (`https://arxiv.org/abs/1912.04958`).

We select the model name that we want to test:

```
# select model name, in this example we use "stylegan_car512"
model_name = 'stylegan_car512'
```

Next, we need to load the generator of GAN. Please remember that we do not need the discriminator to generate the images. It is only used to help the generator to train on generating the images that we want:

```
# load the pretrained model
generator = load_generator(model_name)
```

Now, we send the code to the generator to sample from the latent space. The code is simply random noise. It is the random noise vector, *z*, which we saw in *Figure 7.1*:

```
codes = torch.randn(num, generator.z_space_dim).cuda()
```

Then, we synthetize the image by sending the noise vector (code):

```
# generate the synthetic image from the code
images = synthesize(generator, gan_type, codes)
```

Now, let us visualize the output of the GAN in *Figure 7.2*:

Figure 7.2 – Images generated using StyleGAN

After changing the latent semantic parameters as described by **SeFa**, we get the outputs shown in *Figure 7.3*:

Figure 7.3 – SeFa approach for controlling generation process by changing latent semantic parameters

In the same way, we can generate images of anime facial expressions, celebrities' faces, and bedroom layouts using the aforementioned models, as shown in *Figure 7.4*:

Figure 7.4 – A sample of images generated using different GAN models

As we have seen in this example, we can effortlessly utilize GANs to generate diverse and photorealistic data for training and testing our own ML models. Next, we will explore the variations of GANs that facilitate many amazing applications.

Variations of GANs

In this section, we will explore the main variation of GANs. For an interesting practical application of GANs, please refer to *Chapter 12* and *Case Study 3 – Predictive Analytics* to see how Amazon utilized GANs for fraud transaction prediction. For more applications, please refer to *Generative Adversarial Networks in the built environment: A comprehensive review of the application of GANs across data types and scales* (`https://www.sciencedirect.com/science/article/abs/pii/S0360132322007089`).

Conditional GAN (cGAN)

A typical GAN generates images given a random noise vector. However, in many scenarios, we really want to control the attributes and properties of the generated synthetic samples. For example, suppose you are deploying a GAN to generate human faces. The standard GAN architecture has no way to let you specify some attributes of the generated faces such as gender, age, eye color, and hair length. Using cGAN, we can condition the GAN on these attributes in the training process. Thus, we are able to generate synthetic samples with certain attributes. For more details, refer to *Conditional Generative Adversarial Nets* at `https://arxiv.org/abs/1411.1784`.

CycleGAN

In an image-to-image translation task that aims to transform an image from one domain to another, DL models usually require matching pairs or pairwise correspondences between images from the two domains. This is extremely difficult to achieve. For instance, imagine preparing such a dataset for mapping images from one season (winter) to another (summer). An elegant solution to the problem is using CycleGANs, which can be trained to perform unpaired image-to-image translation between domains given only two sets of images from both domains without the need for any matching pairs. Thus, you only need to provide images taken in winter and summer and there is no need to capture the same scenes in winter and summer to provide matching pairs. For more details, please check *Unpaired Image-to-Image Translation using Cycle-Consistent Adversarial Networks* (`https://arxiv.org/abs/1703.10593`).

Conditional Tabular GAN (CTGAN)

CTGANs are a specific variant of GANs that can generate tabular synthetic data. It is very challenging for other GANs to capture the dependencies between columns or attributes of a given tabular dataset. A CTGAN is a cGAN that can be utilized to model these joint probability distributions between these columns. CTGANs have enormous applications in data augmentation, imputation, and anomaly detection. For more details, please refer to *Modeling Tabular Data using Conditional GAN* (`https://arxiv.org/abs/1907.00503`).

Wasserstein GAN (WGAN) and Wasserstein GAN with Gradient Penalty (WGAN-GP)

WGAN and WGAN-GP are variants of the original GANs. Unlike GANs, which use a binary cross-entropy loss to classify real and fake samples, this variation utilizes Wasserstein distance to measure the distance between the real and fake data probability distributions. Furthermore, WGAN-GP implements a gradient penalty term to enforce the Lipschitz constraint on the discriminator. These two variants were shown to produce better results and to be more stable. For more details, check *Wasserstein GAN* (`https://arxiv.org/abs/1701.07875`) and *Improved Training of Wasserstein GANs* (`https://arxiv.org/abs/1704.00028`).

f-GAN

f-GANs are another family of GANs that utilize *f*-divergences to measure and minimize the divergence between real and fake samples' probability distributions. This variant of GANs has been widely utilized in image and text generation. For more details, please check *f-GAN: Training Generative Neural Samplers using Variational Divergence Minimization* (`https://arxiv.org/abs/1606.00709`).

DragGAN

DragGANs are another recent promising variation of GANs that open the door for many amazing applications, such as point-based image editing. DragGANs allow users to generate photorealistic synthetic images in an interactive and intuitive manner. DragGANs stand out due to their distinctive approach to optimizing the latent space and their unique method of point tracking. For more information, please refer to *Drag Your GAN: Interactive Point-based Manipulation on the Generative Image Manifold* (https://arxiv.org/abs/2305.10973).

Let's wrap things up before we move on.

Summary

In this chapter, we have discussed what GANs are, their architecture, and the training process. At the same time, we explored how GANs were utilized for various applications such as image-to-image translation. Additionally, we covered a coding example demonstrating how to use GANs to generate photorealistic images. In this chapter, we also learned about the main variations of GANs. In the next chapter, we will continue our learning journey by exploring another exciting approach for generating synthetic data by utilizing video games.

8

Video Games as a Source of Synthetic Data

In this chapter, you will learn how and why to use video games for synthetic data generation. We will discuss the current research and highlight the resultant advancements made in this area. At the same time, we will delve into the main challenges of utilizing this approach for synthetic data generation. The overall aim of this chapter is to help you gain a thorough understanding of utilizing video games for synthetic data generation. It will equip you with the necessary knowledge to understand the potentialities and limitations of this promising synthetic data generation approach.

In this chapter, we're going to cover the following main topics:

- Video games and synthetic data
- Generating synthetic data using video games
- Challenges and limitations

The impact of the video game industry

In this section, we will understand why video games present an ideal medium for synthetic data generation.

Video games are interactive electronic games used primarily for entertainment. The player usually interacts with game elements to achieve an objective. The volume, quality, and quantity of games released each year have grown exponentially in recent years. Video games are now utilized in education, training, rehabilitation, personal development, and just recently in **machine learning** (**ML**) research. Specifically, they are presented as a rich and excellent synthetic data resource for training and testing ML models.

Synthetic data generation approaches such as simulators and GANs are promising and present a clever solution for generating large-scale and automatically annotated datasets; refer to *Chapters 6* and *7*. However, they still have a few limitations due to the complexity, time, and effort required to set up the data generation system. Some of these limitations can be summarized in the following:

- Photorealism and domain shift

- Time, effort, and cost

Next, let's discuss these limitations one by one in more detail.

Photorealism and the real-synthetic domain shift

For synthetic data to be useful, most often, it needs to be photorealistic. However, attaining a high level of photorealism is not simple. If you recall, we briefly discussed these points in *Chapter 6*. It requires skilled engineers, artists, designers, experts in lighting and animation, and others to achieve this aim. In addition, the hardware to run these state-of-the-art game engines, the game engine itself, the tools, and the assets to build a rich and realistic virtual world are expensive and require extensive training. For example, the *Wacom Cintiq Pro 24"* costs more than £2,200 and requires months of training. This and similar devices are essential to create 3D content for games. Please refer to *Figure 8.1*.

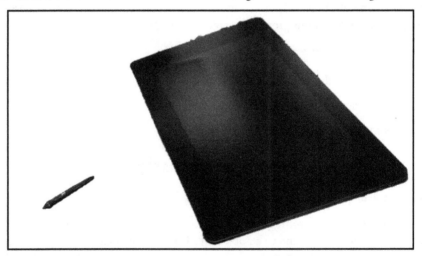

Figure 8.1 – An example of a tool needed for 3D or 2D workflows in games

The following are the system requirements to run Unreal Engine:

- Windows 10 64-bit, quad-core Intel or AMD, 2.5 GHz or faster, 8 GB RAM

- macOS Big Sur, quad-core Intel, 2.5 GHz or faster, 8 GB RAM

- Linux Ubuntu 18.04, quad-core Intel or AMD, 2.5 GHz or faster, 32 GB RAM

Budgeting for the software, hardware, and experts to create a system for synthetic data generation may not be feasible for small-sized companies and research groups. Failing to maintain the required photorealism causes ML models to struggle at generalizing from synthetic to real data. Thus, it limits the usability of the generated synthetic data in real-world applications.

Time, effort, and cost

Creating the multi-purpose simulator *CARLA* (`https://carla.org`) required 40-50 researchers, developers, and artists. It took several years of consistent work to develop this useful system for synthetic data generation. Similarly, GANs require a considerable amount of time and effort to design and train. Additionally, significant computational and data storage resources are required for the training. Ethical considerations are other issues with this approach of synthetic data generation. Intellectual property issues and bias are examples of these concerns.

Searching for an ideal synthetic data generation approach is still a hot topic and a problem that is partially solved. The recent advancements in the game industry motivated researchers to explore the potential of using video games to generate synthetic data. The number of users in the video game industry is expected to reach 3.1 billion by 2027. The projected market volume is expected to achieve more than £380 billion by 2027. For more details, refer to *Statista Video Games Worldwide* (`https://www.statista.com/outlook/dmo/digital-media/video-games/worldwide`).

The global market, the great advancement in technology, and the increasing demand for more photorealistic and engaging games opened the doors for researchers, game developers, game studios, investors, and game publishers to push the boundaries for creating unprecedented and ground-breaking photorealistic and diverse virtual worlds and games. Now, you understand why video games present an elegant solution to generating synthetic data. In the next section, we will go into more detail on how to utilize video games for synthetic data generation.

Generating synthetic data using video games

This synthetic data generation approach transfers the problem from creating virtual worlds to generate synthetic data to manipulating a video game to generate synthetic data instead. This method presents a convenient and efficient way to generate synthetic data. Examples of video games that have been leveraged for this purpose are listed as follows:

- Grand Theft Auto V: *Playing for Data: Ground Truth from Computer Games* (`https://arxiv.org/pdf/1608.02192v1.pdf`)

- Minecraft: *Exploring the Impacts from Datasets to Monocular Depth Estimation (MDE) Models with MineNavi* (`https://arxiv.org/pdf/2008.08454.pdf`)

- Half-Life 2: *OVVV: Using Virtual Worlds to Design and Evaluate Surveillance Systems* (`https://www.computer.org/csdl/proceedings-article/cvpr/2007/04270516/12OmNyRg4Dv`)

The following diagram shows the varied genres of video games:

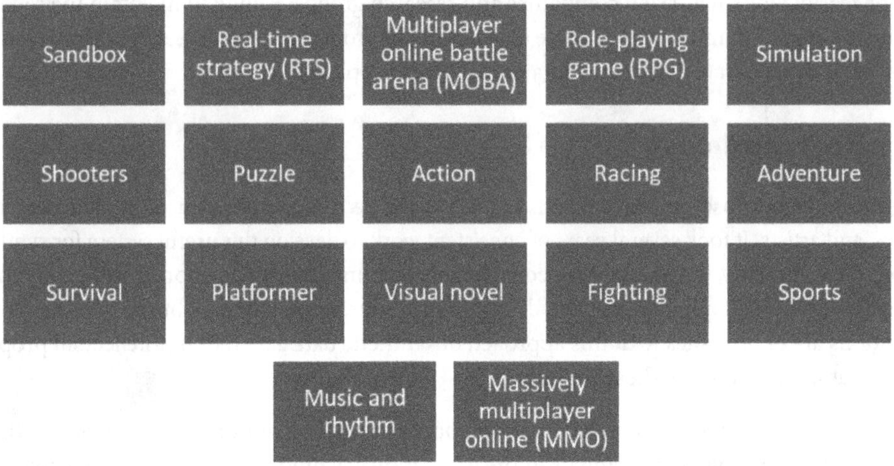

Sandbox	Real-time strategy (RTS)	Multiplayer online battle arena (MOBA)	Role-playing game (RPG)	Simulation
Shooters	Puzzle	Action	Racing	Adventure
Survival	Platformer	Visual novel	Fighting	Sports
	Music and rhythm	Massively multiplayer online (MMO)		

Figure 8.2 – Examples of video game genres

The diversity in video game genres, as shown in *Figure 8.2*, makes generating various and rich synthetic data by utilizing video games a more attractive option for researchers. There are various ways to leverage video games in ML applications, such as the following:

- Utilizing games for general data collection
- Utilizing games for social studies
- Utilizing simulation games for data generation

Let's discuss these in detail.

Utilizing games for general data collection

Modern video games collect data about players for various reasons, such as controlling the game difficulty, customizing the game experience, and personalizing advertisements based on users' data. The collected data by these games can be leveraged for other purposes beyond the game itself. The data collected can include the following:

- Biometric data
- Demographic data
- Gameplay data

Let's see how this data can be used.

Biometric data

Game developers may collect players' biometric data to assess their engagement and may control certain aspects of the game based on the player's emotions, excitement, and confusion. The biometric data collected by some modern video games can include eye movements, heart rate, head and hand movements, and **electroencephalography** (**EEG**) signals. EEG signals are electrical waves generated by the brain when performing activities such as attention, reading, and sleeping. This includes video games such as *Nevermind, Far Cry 6*, and *X-Plane 12. Nevermind* (https://nevermindgame. com) is the first physiological biofeedback adventure video game. It captures the player's heart rate using a heart rate sensor to assess fear, anxiety, and stress during playtime. The collected data can be utilized for a wide range of experiments, such as *Video Games and Stress: How Stress Appraisals and Game Content Affect Cardiovascular and Emotion Outcomes* (https://www.frontiersin.org/ articles/10.3389/fpsyg.2019.00967/full) and *Interaction Effects of Action Real-Time Strategy Game Experience and Trait Anxiety on Brain Functions Measured via EEG Rhythm* (https:// www.tandfonline.com/doi/full/10.1080/27706710.2023.2176004).

Demographic data

Video games can be considered a rich source of demographic data. Widespread global players provide valuable information about their age, gender, location, language, occupation, and other valuable information such as **Personal Identifiable Information** (**PII**). Some video games, such as *Minecraft* and *The Sims 4,* utilize the collected demographic data to improve the player experience and enhance engagement. Thus, some video games may remove violence, blood, and mature language for players under 7 years old. They may activate or deactivate certain features based on the demographic data collected, such as in-game chat and purchases and maturity filters.

Gameplay data

Games may collect data explicitly by asking questions to users or by using sensors, and implicitly by observing the behavior of the player in the virtual world including interaction, choices, learning, and progress. Gameplay data provides valuable resources to customize your game and improve the overall gameplay and user experience. Examples of these games are *Call of Duty: Black Ops III, Fortnite, League of Legends*, and *World of Warcraft*. The game developers leverage the collected data during playtime to balance gameplay mechanics and difficulty and identify weak features, issues, and bugs.

Utilizing games for social studies

Games have been utilized in the field of social studies as a source of synthetic data. For example, educational games were deployed by researchers to understand more about learning experiences, compare and contrast teaching methods, and analyze learning environments and learning tools. For more information about utilizing games in social studies, refer to *Teaching Social Studies with Games* (https://www.igi-global.com/article/teaching-social-studies- with-games/201873).

In some games, such as *Civilization V* (`https://civilization.com/en-GB/civilization-5`), the players build and control a civilization. They need to make decisions, plan, conduct diplomacy, and make plans for the future. Observing and recording the players' behaviors and combining it with the demographic data collected from the players present an excellent and rich source of data for social studies researchers. Other similar games, such as *Spent* (`https://playspent.org`), also provide valuable synthetic data on how players respond to challenges and how they make decisions with a limited budget.

As expected, the standard methods of collecting data for social studies are surveys and interviews. They are hard to design, time-consuming, and expensive. In addition to that, they are not pleasant and enjoyable to do. Thus, wrong or inaccurate feedback or data is expected. On the other hand, games present an interactive, realistic, and enjoyable experience, which makes the feedback and data generated more accurate, detailed, and valuable.

Utilizing simulation games for data generation

Simulation games create parts of the real world that can be leveraged to teach players something, equip them with new skills, or sharpen their knowledge and skills. Simulators generate synthetic data from two sources: the environment and the behavior of the player in the simulation game. The simulated environment gives valuable information such as semantic labels of the objects, weather conditions, and scene crowdedness. On the other hand, the behavior of the player includes how the player approaches the problem, time spent, resources utilized, decisions made, and communication with other players in the game.

As you now understand, the collected data inside the video games can be utilized for many purposes such as training ML models. However, game developers must consider data regulations to ensure they do not violate or breach them. In the next section, you will learn about the main limitations of this synthetic data generation approach.

Challenges and limitations

In this section, we explore the main challenges of this synthetic data generation approach.

Similar to other synthetic data generation approaches, utilizing video games for this purpose has some challenges. The main challenges and drawbacks can be summarized as follows.

Controllability

One of the main advantages of using synthetic data is the ability to customize and fully control each aspect of this process. However, with video games not deployed for research purposes, the 3D virtual world and the scene elements are already created, fixed, and hard to change. Thus, it is not easy to add or remove certain objects and actions.

Game genres and limitations on synthetic data generation

Depending on the game market, some game genres are more popular than others. For example, *action*, *sports*, *role-playing*, and *first-person shooter* games are more popular compared to *text-based adventure* and *music* and *rhythm*. Thus, generating synthetic data for certain applications may not be feasible given the limited availability of suitable video games.

Realism

As we have mentioned earlier, realism is important to bridge the gap between synthetic and real domains. Many state-of-the-art and best-selling games are far from being realistic games. Examples of these games are *PUBG*, *Minecraft*, and *Red Dead Redemption*. Players usually value gameplay, plot, and narrative more than photorealism in games. Achieving realistic graphics in video games is an extremely challenging process and may not be rewarding to games studios.

Ethical issues

Utilizing a video game for synthetic data generation is a complex topic. Using the players' data outside of the game for research or commercial projects may cause privacy issues. Training on synthetic data generated from video games will inherently make ML models trained on this data source more accurate at demographic groups usually playing video games and less with other groups who do not. For example, a face recognition system trained exclusively on synthetic data generated by video games may perform well on young adults but may struggle with older adults who do not usually play video games. Please refer to *Distribution of video gamers in the United States in 2022, by age group* (`https://www.statista.com/statistics/189582/age-of-us-video-game-players`).

Intellectual property

We cannot utilize video games for synthetic data generation without permission from the creators of the game, as the creators of the game usually have exclusive rights to reproduce the game. Intellectual properties and copyrights may be violated if we modify the game and utilize it for other aims. At the same time, the data generated from the game itself might be subject to the same copyrights or license.

Summary

In this chapter, we have learned how video games present a valuable source of synthetic data. We learned that biometric, demographic, and gameplay data can be collected from video games and can be utilized for myriads of ML-based applications. We gained insight into how and why various types of data about the player can be collected from video games. Finally, we discussed the main limitations of using video games for synthetic data generation.

In the next chapter, we will delve into another method for generating synthetic data, using diffusion models.

9

Exploring Diffusion Models
for Synthetic Data

This chapter introduces you to diffusion models, which are cutting-edge approaches to synthetic data generation. We will highlight the pros and cons of this novel synthetic data generation approach. This will help you to make informed decisions about the best methods to utilize for your own problems. We will highlight the opportunities and challenges of diffusion models. Moreover, this chapter is enriched with a comprehensive practical example, providing hands-on experience in both generating and effectively employing synthetic data for a real-world ML application. As you go through diffusion models, you will learn about the main ethical issues and concerns around utilizing this synthetic data approach in practice. In addition to that, we will review some state-of-the-art research on this topic. Thus, this chapter will equip you with the necessary knowledge to thoroughly understand this novel synthetic data generation approach.

In this chapter, we will cover the following main topics:

- An introduction to diffusion models

- Diffusion models – the pros and cons

- Hands-on diffusion models in practice

- Diffusion models – ethical issues

Technical requirements

Any code used in this chapter will be available under the corresponding chapter folder at this book's GitHub repository: `https://github.com/PacktPublishing/Synthetic-Data-for-Machine-Learning`.

An introduction to diffusion models

In this section, we will explore diffusion models. We will compare them to **Variational Autoencoders** (**VAEs**) and **Generative Adversarial Networks** (**GANs**), which we covered in *Chapter 7*. This will help you to gain a holistic and comprehensive understanding of generative models. Additionally, it will make comparing and contrasting the architectures, training procedures, and data flow of these methods straightforward. Furthermore, we will also learn how to train a typical diffusion model.

Diffusion Models (**DMs**) are generative models that were recently proposed as a clever solution to generate images, audio, videos, time series, and texts. DMs are excellent at modeling complex probability distributions, structures, temporal dependencies, and correlations in data. The initial mathematical model behind DMs was first proposed and applied in the field of statistical mechanics to study the random motion of particles in gases and liquids. As we will see later, it is essential and crucial to learn about DMs, as they are powerful generative models that can usually generate higher-quality and more privacy-preserving synthetic data compared to other approaches. Additionally, DMs rely on strong mathematical and theoretical foundations.

One of the first works to show that DMs can be utilized to generate photorealistic images was *Denoising Diffusion Probabilistic Models* (`https://arxiv.org/abs/2006.11239`), which was proposed by researchers from UC Berkeley. This pioneering work was followed by another work by OpenAI titled *Diffusion Models Beat GANs on Image Synthesis* (`https://arxiv.org/pdf/2105.05233.pdf`), showing that DMs are better at generating photorealistic synthetic images. Then, other researchers started to explore the potential of these DMs in different fields and compare them to VAEs and GANs.

Variational Autoencoders (**VAEs**) are one of the earliest solutions for generating synthetic data. They are based on using an encoder to encode data from a high-dimensional space (such as RGB images) into a latent low-dimensional space. Then, the decoder is used to reconstruct these encoded samples from the latent space to the original high-dimensional space. In the training process, the VAE is forced to minimize the loss between the original training sample and the reconstructed one by the decoder. Assuming the model was trained on a sufficient number of training samples, it can then be used to generate new synthetic data by sampling points from the latent space and using the decoder to decode them, from the latent low-dimensional space to the high-dimensional one, as shown in *Figure 9.1*.

The training process of DMs

DMs are used to generate synthetic data with a distribution close to the probability distribution of the training data. Thus, an important task is to learn the distribution of our training data and then leverage our knowledge of the real data to generate an unlimited number of synthetic data samples. Usually, we would want to generate high-quality and diverse synthetic data using a fast-generation method. However, each generation method has its own advantages and disadvantages, as we will see later on.

As illustrated in *Figure 9.1*, given a training image, x_0, from the real data, the DM adds **Gaussian noise** to this image to become x_1. The process is repeated until the image simply becomes an image of random noise, z. This process is called **forward diffusion**. Following this, the model starts the denoising process, in which the DM takes the random noise, z, and reverses the previous process (i.e., forward diffusion) to reconstruct the training image. This process is known as **reverse diffusion**.

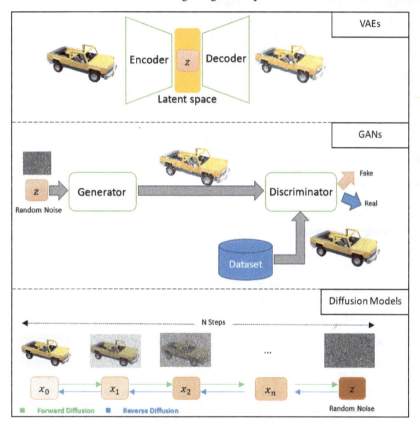

Figure 9.1 – The training process and architectures of the main generative models – VAEs, GANs, and DMs

As we can see from *Figure 9.1*, the forward diffusion process is a **Markov chain**. Each step of the process is a stochastic event, and each event depends only on the previous event or state. Thus, they form a sequence of stochastic events and, consequently, a Markov chain.

A key idea of the DMs' training process is using a neural network such as *U-Net* (`https://arxiv.org/abs/1505.04597`) to predict the amount of noise that was added to a given noisy image from the training data. This is crucial to reverse the noising process and learn how to generate a synthetic sample given random noise.

When the diffusion model converges after a successful training process, we can give it random noise, z, and the DM, using the reverse diffusion path, will give us a synthetic sample based on the provided z. Thus, we can now generate an unlimited number of new synthetic samples from the same training data probability distribution.

Please note the number of diffusion steps, N, in the noising/denoising process depends on how smooth you want the training process to be. In other words, a higher value of N means less abrupt and more gradual noise will be added in the training process. Thus, the model's weights will update steadily and the optimization loss will decrease smoothly. However, higher values of N will make the process slower (usually, $N = 1,000$).

Applications of DMs

DMs have been utilized for a wide range of applications in various domains such as computer vision, natural language processing, finance, and healthcare. Let's briefly discuss some of these applications.

- **Computer vision:**

  - **Image generation**: Generating diverse photorealistic images is one of the ultimate aims of DMs. However, there is usually a trade-off between diversity and photorealism. Thus, to maintain good fidelity, the DMs are usually guided or conditioned on textual data. For more information, please refer to *GLIDE: Towards Photorealistic Image Generation and Editing with Text-Guided Diffusion Models* (`https://arxiv.org/abs/2112.10741`). As you can imagine, the generated synthetic images can be used for various applications, including data augmentation, creative art, prototyping, and visualization.

  - **Video prediction**: Predicting the next frame has many useful applications, as it facilitates predicting future events, which is essential for planning and decision-making in fields such as robotics. Simultaneously, it has huge applications in fields such as surveillance and security. Video prediction ML models can be utilized to anticipate and forecast possible threats, hazards, and potential risks. Additionally, predicting future frames accurately can be utilized to generate synthetic videos to complement real training data. For more details, refer to *Diffusion Models for Video Prediction and Infilling* (`https://arxiv.org/pdf/2206.07696.pdf`), *Video Diffusion Models* (`https://arxiv.org/pdf/2204.03458.pdf`), and *Imagen Video: High-Definition Video Generation with Diffusion Models* (`https://arxiv.org/pdf/2210.02303.pdf`).

- **Image inpainting**: This is simply filling in or restoring missing or damaged parts in images. It is a fundamental task in areas such as historical archiving, privacy protection, and entertainment. Furthermore, it has also been recently utilized for synthetic data generation. For example, DMs were utilized recently to generate synthetic brain MRIs, with the ability to control tumoral and non-tumoral tissues. It was shown that using generated synthetic data can boost performance considerably. For more details, please refer to *Multitask Brain Tumor Inpainting with Diffusion Models: A Methodological Report* (`https://arxiv.org/ftp/arxiv/papers/2210/2210.12113.pdf`) and *RePaint: Inpainting using Denoising Diffusion Probabilistic Models* (`https://arxiv.org/abs/2201.09865`).

- **Image colorization**: This is the task of transferring grayscale images into colored ones. For example, this is essential to improve the photorealism of historical or old photographs. Adding color is important to make these photos more appealing and more emotionally engaging. DMs were shown to perform very well under this task. For further information, please read *Palette: Image-to-Image Diffusion Models* (`https://arxiv.org/pdf/2111.05826.pdf`).

- **Natural language processing**:

  - **Text generation**: Virtual assistants, chatbots, and similar conversational text-based systems rely on text generation. DMs have been utilized recently for this task to improve the quality and diversity, as they are more capable of capturing complex distributions. For an example, please refer to *DiffuSeq: Sequence to Sequence Text Generation with Diffusion Models* (`https://arxiv.org/abs/2210.08933`).

  - **Text-to-speech synthesis**: This is the process of transforming text into audio. While it has many applications in fields such as **Human-Computer Interaction** (**HCI**), education and learning, and video games, it has been recently utilized to make textual content accessible to individuals with visual impairment. For more details about DM and text-to-speech synthesis, please refer to *Diff-TTS: A Denoising Diffusion Model for Text-to-Speech* (`https://arxiv.org/pdf/2104.01409.pdf`) and *Prodiff: Progressive fast diffusion model for high-quality text-to-speech* (`https://dl.acm.org/doi/abs/10.1145/3503161.3547855`). Additionally, for a survey of recent text-to-speech DM-based methods, please refer to *A Survey on Audio Diffusion Models: Text to Speech Synthesis and Enhancement in Generative AI* (`https://arxiv.org/pdf/2303.13336.pdf`).

  - **Text-driven image generation**: This is another promising field where the aim is to generate visual content, such as images based on textual input. It has various applications in content generation, marketing, data augmentation, and assisted data-labeling tools. As expected, DMs are excellent at modeling complex data distribution and very powerful at generating diverse and appealing images. Thus, they have been utilized for text-driven image generation. For more details, please refer to *Text2Human: text-driven controllable human image generation* (`https://dl.acm.org/doi/abs/10.1145/3528223.3530104`).

- **Other applications:**

  - **Privacy in healthcare**: Real **Electronic Health Records** (**EHRs**) of patients carry rich and very useful information that ML models can leverage to help in disease diagnosis, predictive analytics, decision-making, and the optimization and management of resources. DMs were shown to generate high-quality and large-scale EHRs that can be leveraged in enormous applications, such as ML model training, healthcare research, and medical education. To delve into more details, please read *MedDiff: Generating Electronic Health Records using Accelerated Denoising Diffusion Model* (`https://arxiv.org/pdf/2302.04355.pdf`).

  - **Anomaly detection**: This is the task of detecting or identifying patterns and instances that are not in line with the expected behavior and distribution. It has a myriad of applications in cybersecurity, fraud detection, telecommunication, and manufacturing. DMs are usually robust to noise and more stable, which makes them ideal for these applications. For an example and more details about utilizing DMs for anomaly detection in healthcare, please refer to *Diffusion models for medical anomaly detection* (`https://link.springer.com/chapter/10.1007/978-3-031-16452-1_4`).

  - **Text-to-motion**: Generating animation or motion from textual input has many applications in the training, education, media, and entertainment sectors. DMs have shown promising results by producing high-quality animations of human motion. For more details, please refer to *Human Motion Diffusion Model* (`https://arxiv.org/abs/2209.14916`).

Now that we have an idea of the domains where diffusion models are used, in the next section, we closely examine the pros and cons of DMs.

Diffusion models – the pros and cons

In this section, you will learn about and examine the main pros and cons of using DMs for synthetic data generation. This will help you to weigh the advantages and disadvantages of each synthetic data generation method. Consequently, it will give you the wisdom to select the best approach for your own problems.

As we learned in *Chapter 7*, GANs work very well for certain applications, such as style transfer and image-to-image translation, but they are usually very hard to train and unstable. Additionally, the generated synthetic samples are usually less diverse and photorealistic. Conversely, recent papers have shown that DM-based synthetic data generation approaches surpass GANs on many benchmarks. For more details, please refer to *Diffusion Models Beat GANs on Image Synthesis* (`https://arxiv.org/pdf/2105.05233.pdf`). Like any other synthetic data generation approach, DMs have pros and cons. Thus, you need to consider them carefully for your particular application or problem. Then, you can select the best approach to generate the synthetic data that you want. With that aim in mind, we will examine the key advantages and disadvantages of using DMs.

The pros of using DMs

In general, DMs are excellent at modeling complex data probability distributions and capturing temporal dependencies and hidden patterns. This is possible with DMs because they use a diffusion process to model data distributions, using a sequence of conditional distributions. Thus, we can highlight the main strengths and merits of DMs as follows:

- **Generalizability and applicability to a wide range of problems**: Unlike other generative methods, which are limited to image and video generation, DMs can be utilized to generate images, audio, videos, texts, molecular structures, and many other data types and modalities

- **Stability in the training process**: The architecture, the training process, and the optimization technique of DMs make them more stable compared to other generative models such as GANs

- **High-quality synthetic data generation**: Due to their distinctive architecture and innovative gradual and iterative training process, DMs generate high-quality synthetic data that surpasses other generative models such as VAEs and GANs

The cons of using DMS

The two main limitations and shortcomings of using DMs can be described in terms of the computational complexity of the training and inference process, as well as the large-scale training datasets required by DMs:

- **Computational complexity**: DMs are computationally heavy. They are usually slow compared to other generative models, as the forward and reverse diffusion processes are composed of hundreds of steps (usually, the number of steps, N, is close to 1,000).

- **More training data is required**: DMs require large-scale training datasets to converge. Obtaining such datasets is not suitable for certain fields, which limits the usability of DMs for certain applications.

Now, we can clearly identify the pros and cons of using DMs for synthetic data generation. Let's practice utilizing DMs to generate synthetic data to train ML models.

Hands-on diffusion models in practice

Let's study a practical example that demonstrates the usability of synthetic data in the computer vision field. For that aim, we will generate and prepare our dataset, build our ML model from scratch, train it, and evaluate its performance. The dataset is available at *Kaggle* (`https://www.kaggle.com/datasets/abdulrahmankerim/crash-car-image-hybrid-dataset-ccih`). The full code, the trained model, and the results are available on GitHub under the corresponding chapter folder in the book's repository.

Context

We want to build an ML model that can classify car images into two distinct categories – images depicting car accidents and those that do not. As you can imagine, curating such a real dataset is time-consuming and error-prone. It could be easy to collect car images without accidents. However, collecting images of cars with accidents, collisions, fires, and other dangerous scenarios is extremely hard. To solve this problem and to prove the usability of synthetic data, let's first generate our training dataset. We can use a single synthetic data generation approach for that aim. However, let's combine different methods and tools to collect more diverse data and practice different approaches as well. In this example, we will use the following methods:

- The DALL·E 2 image generator
- The DeepAI text-to-image generator
- A simulator built using a game engine such as Silver

Dataset

First, let's leverage the remarkable capabilities of recent generative models such as *DALL·E 2* (`https://openai.com/dall-e-2`) to generate some car accident images.

Figure 9.2 – Generating synthetic images using the DALL·E 2 web application

We can simply use the following prompts to generate these images (see *Figure 9.2*):

- `Car accidents`
- `White car accidents`
- `Red car accidents`
- `Blue car accidents`
- `Ambulance car accident`

As you can see from *Figure 9.3*, the generated images look photorealistic and diverse, which is exactly what we need to train our ML model.

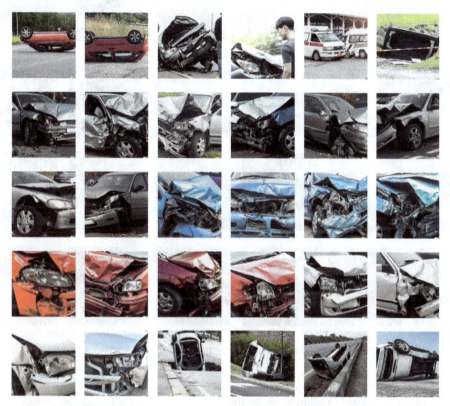

Figure 9.3 – Car accident images generated using DALL·E 2

We also use images collected from the *car-accident(resnet)* dataset, licensed under CC BY 4.0 (`https://universe.roboflow.com/resnet-car-accident/car-accident-resnet-n7jei`) using Roboflow (`https://roboflow.com`). We chose this dataset as the images are similar to what they would be if they were sourced from a video game, such as *BeamNG Drive Crashes* (`https://www.beamng.com/game`). We will add these images to our dataset to further improve its diversity. A sample of these images is shown in *Figure 9.4*.

Figure 9.4 – Car accident images collected from a video game

Then, we need to generate similar images for the other category, images of cars with no accidents. This time, let's use the *DeepAI* tool to generate these images. As did earlier, we can simply use the following prompts to get the required images:

- `Car under rain`
- `White car`
- `Car under fog`
- `Blue car`

As we can see in *Figure 9.5*, we effortlessly obtained another 30 car images with no accidents.

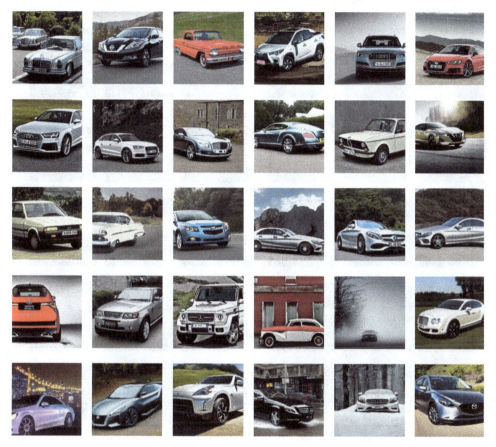

Figure 9.5 – Intact car images generated using DeepAI

Now, we have 60 synthetic images, but we still need more images to appropriately train our ML model. We can generate any number of images using the previous generative models, but let's explore another way – using the *Silver* simulator (https://github.com/lsmcolab/Silver).

By specifying the number of images that we want, we can generate the following images for this category (*Figure 9.6*).

Figure 9.6 – Intact car images generated using Silver

At this point, we have collected 600 synthetic images. Now, to assess the performance of our trained model, we should test it on real car images. Let's collect real images using the *Unsplash* website (https://unsplash.com). To further improve our dataset, let's also manually add images using Roboflow. *Figure 9.7* shows a sample of these images.

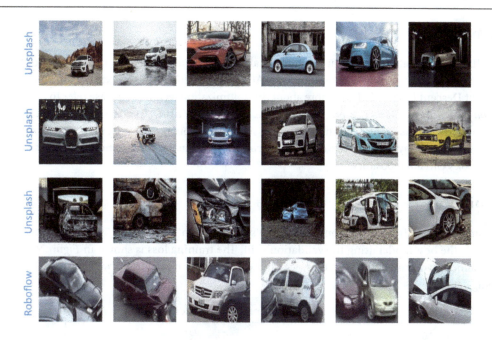

Figure 9.7 – Sample images of our real dataset

Finally, our dataset is composed of 600 synthetic images and 250 real ones, as shown in *Table 9.1*.

Split	Synthetic	Real
Training	540	-
Validation	60	62
Test	-	188
Total	600	250

Table 9.1 – Our final dataset's splits and number of images

Note that we will train only on synthetic data and test only on real data. Also, note in the validation split that we used both synthetic and real data because we are training and testing on two different domains – synthetic and real. Thus, a balanced mixture of data from both domains is necessary to give a good understanding of our model's learning of the synthetic data and generalizability to the real domain.

ML model

Our ML model is composed of four convolutional and three fully connected layers, max pooling, dropout, and batch normalization. We will use the **Adam** optimizer, which is an extension of **Stochastic Gradient Descent (SGD)**. To learn more about this optimization technique, please refer to *Adam: A Method for Stochastic Optimization* (https://arxiv.org/abs/1412.6980). Additionally, since the problem (as mentioned in the *Context* section) is a binary classification, we will deploy **Binary Cross Entropy (BCE)** loss. Refer to the train.py file in the corresponding chapter's folder of the book's GitHub repository.

Training

We train our model from scratch on the training split shown in *Table 9.1* for 30 epochs. Then, we will select the best model using the validation split. The training loss is shown in *Figure 9.8*. The loss is smoothly decreasing, as expected, which means that our model trains well on our synthetic training dataset.

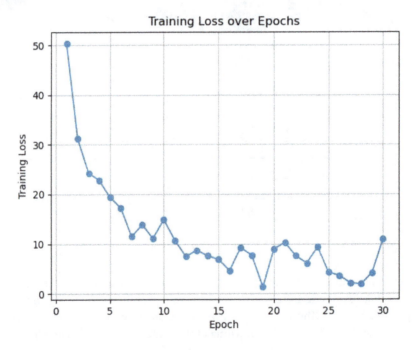

Figure 9.8 – The training loss during the training stage

From the validation accuracy shown in *Figure 9.9*, we can see that our model achieved the best results on the validation set at epoch 27. Thus, we will use this model for testing.

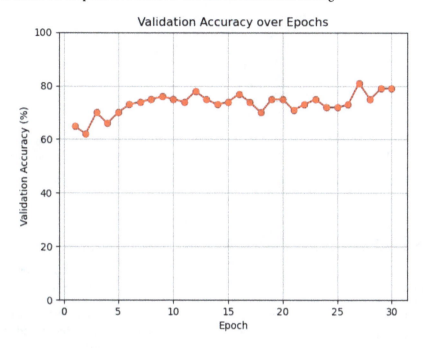

Figure 9.9 – Validation accuracy during the training stage

Testing

Let's see the performance of our best model on the test split. As you can see from the confusion matrix shown in *Figure 9.10*, our model classifies no-accident cars with an accuracy of *84%*, while it classifies accidents with an accuracy of *74%*. Our model achieved a total accuracy of *80%*, which is excellent given that our model was trained only on synthetic data.

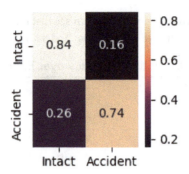

Figure 9.10 – The confusion matrix illustrating classification results

We have provided a detailed example of how to generate and leverage synthetic data for training ML models. Now, let's delve into DMs and the main ethical issues.

Diffusion models – ethical issues

In this section, you will learn about the main ethical issues associated with using DMs for synthetic data generation.

Diffusion-based generative models are emerging and powerful technologies. Thus, their pros and cons need to be considered carefully. Their advantages are huge for businesses, industry, and research. However, they possess dangerous capabilities that can be leveraged to cause harm to individuals, businesses, societies, and so on.

Let's list the main ethical issues usually associated with generative models and, especially, DMs:

- Copyright
- Bias
- Inappropriate content
- Responsibility issues
- Privacy issues
- Fraud and identity theft

Now, let's delve into some of the main ethical issues behind using DMs in practice.

Copyright

DMs are usually trained on large-scale real datasets. For example, DALL E 2 was trained on more than 650 million text-image pairs. Thus, obtaining permission from the owners of these data samples, such as images, artworks, video clips, and others, is not feasible. Additionally, DMs may misrepresent or represent copyrighted and intellectual properties with minor changes. For example, *Universal Music* asked streaming services and companies to prevent AI/ML companies from accessing their songs for training (`https://www.billboard.com/pro/universal-music-asks-spotify-apple-stop-ai-access-songs`). Thus, it creates many difficult questions for regulators to address and regulations for companies to comply with. Taking into account the huge, rapid progress in this field, the regulations may not be able to provide an ethical frame to control copyright issues.

Bias

As we have outlined in this chapter, DMs learn to generate synthetic data following the same distribution of the training data. Thus, if the training data is biased, the DM will also generate biased synthetic data. What is more difficult with these complex models is assessing this bias compared to examining the raw training data, such as images and videos. It becomes even more complex and severe when these models are leveraged in decision-making by non-experts. For example, a professional comic artist may identify gender, political, and age biases within a comic book. However, training a DM on millions of comic books and then assessing the biases of the generated comic books, using this model, may not be possible!

Inappropriate content

DMs may generate inappropriate content without users' consent. It may include inappropriate language, sexual and violent content, hate speech, and racism. Thus, DMs need to utilize an effective filtering mechanism to discard unsuitable data. Simultaneously, they need to utilize a safeguarding procedure to prevent generating inappropriate content.

Responsibility

Utilizing generative models such as DMs in decision-making will be almost unavoidable in the future. However, it is not possible, with current DMs, to understand why a certain decision was made. Thus, it is challenging to understand who is responsible for wrong decisions made by DMs that may cause death, injuries, or damage to properties. Thus, we need to develop suitable mechanisms to ensure transparency, accountability, and tracking in the process of decision-making.

Privacy

DMs may reveal sensitive information about individuals and organizations since the generated synthetic data still follows the statistical properties of the original real training data. These models may disclose information that could be utilized by a third party to cause harm, loss, or unwanted consequences. For example, ChatGPT was banned in Italy due to privacy issues.

Fraud and identity theft

DMs are powerful and capable of mimicking the human voice, photos, and videos. Thus, generated fake media can be exploited for many fraudulent purposes, such as accessing personal information, money laundering, credit card fraud, and cybercrime. Furthermore, DMs can be used to impersonate a celebrity, activist, or politician to get access to private information, classified material, and documents.

Summary

In this chapter, we introduced a novel and powerful method to generate synthetic data – using DMs. We compared DMs to other state-of-the-art generative models, and then, we highlighted the training process of DMs. Furthermore, we discussed the pros and cons of utilizing DMs. Additionally, we learned how to generate and utilize synthetic data in practice. We also examined the main ethical considerations usually raised when deploying DMs for synthetic data generation. You developed a comprehensive understanding of generative models, and you learned about standard DM architecture, the training process, and the main advantages, benefits, and limitations of utilizing DMs in practice. In the next chapter, we will shed light on several case studies, highlighting how synthetic data has been successfully utilized to improve computer vision solutions in practice. The chapter aims to inspire and motivate you to explore the potential of synthetic data in your own applications.

Part 4:
Case Studies and
Best Practices

In this part, you will be introduced to rich and diverse case studies in three cutting-edge areas in the field of ML: **Computer Vision (CV)**, **Natural Language Processing (NLP)**, and **Predictive Analytics (PA)**. You will comprehend the benefits of employing synthetic data in these fields and identify the main challenges and issues. Following this, you will learn about best practices that improve the usability of synthetic data in practice.

This part has the following chapters:

- *Chapter 10, Case Study 1 – Computer Vision*
- *Chapter 11, Case Study 2 – Natural Language Processing*
- *Chapter 12, Case Study 3 – Predictive Analytics*
- *Chapter 13, Best Practices for Applying Synthetic Data*

10

Case Study 1 – Computer Vision

In this chapter, you will be introduced to a multitude of industrial applications of computer vision. You will discover some of the key problems that were successfully solved using computer vision. In parallel to this, you will grasp the major issues with traditional computer vision solutions. Additionally, you will explore and comprehend thought-provoking examples of using synthetic data to improve computer vision solutions in practice.

In this chapter, we're going to cover the following main topics:

- Industrial revolutions – computer vision as a solution
- Synthetic data and computer vision – examples from industry

Transforming industries – the power of computer vision

In this section, we'll briefly discuss the main four industrial revolutions as they help us to better comprehend the historical context of AI, data, and computer vision. Then, we will learn why computer vision is becoming an integral component of our modern industries.

The four waves of the industrial revolution

Industrial revolution refers to a global and rapid transformation in the economy. Usually, this transformation brings and utilizes new inventions, discoveries, and technologies to make manufacturing and production processes more efficient. The history of the industrial revolutions can be summarized into four stages:

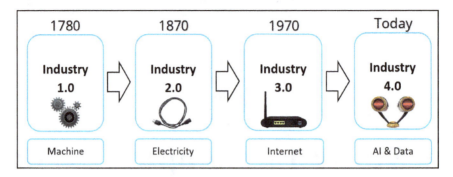

Figure 10.1 – Industrial revolutions

Next, let's discuss each of the industrial revolutions shown in *Figure 10.1* in greater depth.

Industry 1.0

This refers to the first industrial revolution, which happened in the early 19th century. It supplemented, supported, and enhanced existing labor processes by incorporating machinery; animals and manual labor were mostly replaced with water and steam engines. It was a great shift toward using machinery to carry out mostly the same tasks but more efficiently. This opened the door for new industries, such as iron production, which significantly influenced the development of industries such as construction, transportation, and manufacturing. Industry 1.0 changed the way products were produced, which laid the foundation for the next revolution in industry.

Industry 2.0

Electricity was the major driver of the substantial shift in production that happened with Industry 2.0. Assembly line production and the widespread adoption of electricity as a power source facilitated mass production. In parallel to that, the great advancement in steelmaking and production enabled the building of more sophisticated and powerful machinery. This set the stage for the following industrial revolution.

Industry 3.0

Electricity was one of the discoveries that changed our civilization dramatically, including communication and industry. With mass production, which is considered one of the main themes of Industry 2.0, there was an urgent need for automation to minimize errors and maximize efficiency. Thus, computers were utilized by manufacturers to achieve yet more precise and efficient productions.

Industry 4.0

The digital transformations of most industries, great competition between global companies, and scarce resources all opened the door for cyber-physical systems and thus **smart factories**. Consequently, AI, robotics, and **big data** started to attract more attention in industry and academia. Since the main properties of this industrial revolution are great efficiency, customized products, and services, ML and data are the gems of achieving these aims.

Next, we will see why computer vision is the backbone of many of our current industries.

Industry 4.0 and computer vision

Computer vision is an interdisciplinary field that enables machines to understand images. Computer vision is an essential component of our current industrial revolution. It has been widely applied for quality control, safety assurance, predictive maintenance, and other essential and critical applications. Next, let's discuss the main uses of computer vision in the following fields:

- Manufacturing

- Autonomous driving

- Healthcare

- Agriculture

- Surveillance and security

Manufacturing

In manufacturing industries, human error can cause significant delays that affect the entire production pipeline. It may even cause damage to machines and infrastructures, injuries, and death. Computer vision comes as a solution to complement, support, or replace the human element in the process. Computer vision can be utilized to guide the assembly and manufacturing processes to achieve higher throughput with lower costs and higher quality.

Contact lens manufacturers worked with *ADLINK* and *LEDA* that have used computer vision to automate the contact lens inspection process. This task was usually performed by human inspectors where thousands of lenses were visually inspected each day. It was a time-consuming process where errors were not avoidable. By deploying ADLINK and LEDA's computer vision-based solution, the company which manufactures contact lenses was able to make its inspection process 3 times more accurate and 50 times faster! Their novel solution removes the human element from the process, which substantially increases the scalability and quality of the inspection process. For more information about this use case, please refer to *ADLINK and LEDA Technology Create AI-Enabled Contact Lens Inspection Solution* (`https://data.embeddedcomputing.com/uploads/articles/whitepapers/13328.pdf`).

Autonomous driving

This is one of the main sectors that is closely associated with computer vision. **Autonomous driving** technology can reduce human errors in driving and thus minimize accidents, injuries, and death. In 2022, the number of road traffic fatalities exceeded 46,200 cases in the US (https://www.statista.com/statistics/192575/road-traffic-fatalities-in-the-united-states). Thus, computer vision presents a promising safe, efficient, and productive solution. Self-driving companies such as *Tesla*, *Waymo*, and *Mobileye* have already started utilizing computer vision for lane detection and tracking, pedestrian detection, object recognition, and traffic sign detection and recognition. As you may guess, the failure of such computer vision algorithms can cause damage to property, severe injuries, or death. Thus, training and developing a robust computer vision algorithm is a hot topic for ML researchers and is gaining more momentum and receiving more attention.

Tesla cars have developed a computer vision system based on neural networks that takes video inputs from different cameras. Then, it processes the visual information and predicts the road layout, static objects, pedestrians, and other vehicles in the scene. For more information, please refer to *Tesla – AI & Robotics* (https://www.tesla.com/AI).

Aurora Driver is a computer vision system that can be utilized for autonomous driving. The system learns to fuse information collected from various sensors, such as lidar, radar, and cameras, to provide an understanding of the driving environment. For more information, check out *Perception at Aurora: No measurement left behind* (https://blog.aurora.tech/engineering/perception-at-aurora-no-measurement-left-behind).

Next, let's move on to computer vision applications in the healthcare sector.

Healthcare

Computer vision revolutionized the field of healthcare thanks to its great ability to analyze large amounts of patient data and provide quick, accurate, and efficient diagnoses. Computer vision algorithms can assist healthcare practitioners, surgeons, and physicians in making accurate and timely decisions that can reduce costs, improve treatments and operations, and reduce human errors.

A multitude of healthcare providers have already started harnessing computer vision's potential in this field. Let's highlight two examples from *Viz.ai* and *Paige*.

Viz.ai utilizes computer vision algorithms to identify signs of a stroke by analyzing patients' medical images. They deploy these algorithms to efficiently analyze **Computerized Tomography** (**CT**) and **Magnetic Resonance Imaging** (**MRI**) scans and notify neurologists if a sign of a stroke is present to take the appropriate action.

Paige is another company that has deployed computer vision to improve diagnostics and predictive tests of pathologists. In a recent study by *Yale Medicine* on the effectiveness of Paige Prostate (the name of their computer vision tool), 1,876 predictions by this system classified as "suspicious" were reviewed by professional pathologists. The study concluded that only 31.4% of biopsies had to be reviewed by pathologists. Thus, this tool can indeed improve the productivity of pathologists. Additionally, it demonstrated that this tool could improve the detection of prostate cancer especially when being reviewed by non-genitourinary specialized pathologists. For more details, refer to *An independent assessment of AI for prostate cancer detection* (`https://paige.ai/case-study/an-independent-assessment-of-ai-for-prostate-cancer-detection`).

In the following section, let's examine computer vision applications in the agriculture field.

Agriculture

A recent report published by the **Food Security Information Network** (**FSIN**) titled *Global Report on Food Crises 2023* (`https://www.fsinplatform.org/sites/default/files/resources/files/GRFC2023-hi-res.pdf`) raised a red flag about the current and future food insecurity in 58 countries. The report highlighted that almost 250 million people were facing severe food insecurity in 2022, which was a large increase from 2021, when the number was around 190 million. According to experts, the situation is just going to become worse in the future.

Many companies, such as *Taranis* and *Prospera,* utilize computer vision to guide farmers to better optimize resources, analyze crop data, continuously monitor crops, and detect potential issues, such as pests and diseases. Let's talk in more detail about Taranis and Prospera.

Taranis is a company focused on developing technologies that help agriculture businesses and farmers to improve their crop quality, yield, and profit. It utilizes drones and computer vision to analyze farms and make the treatment more efficient. The technology developed is used to efficiently control large farms at the leaf level, which is almost impossible without computer vision. For more information, please refer to the company website (`https://www.taranis.com`).

Prospera is another company that relies on computer vision to support farmers. The technology helps them control pivots, pumps, and other aspects of the irrigation system. Additionally, it continuously monitors crop health and instantly detects any issues. For more details, refer to the company website (`https://prospera.ag`).

As you may expect, these key traditional computer vision solutions can be further enhanced by utilizing synthetic data as a complementary or alternative to real data. Now, let's delve into the main issues with these computer vision solutions, stemming from their significant reliance on real data.

As we have discussed in previous chapters, computer vision algorithms that are based on real data usually suffer from the following issues:

- Insufficient training data
- Data quality issues and bias
- Limited variability

In the next section, you will learn how and why industries have started incorporating synthetic data in their computer vision-based solutions.

Surveillance and security

One of the main key capabilities of computer vision is accurately and efficiently analyzing visual information, such as images and videos. For example, it can be leveraged for the following aims:

- Detecting unusual behaviors

- Identifying suspicious people, items, or actions

- License plate recognition

- Biometric identification: face, iris, palm print, vein, voice, and fingerprint recognition

Thus, computer vision can be leveraged to prevent unauthorized access, protect people and assets, and identify security threats and risks in real time. Computer vision is used these days to ensure public safety. For example, it is used in airports, public transport, streets, parks, and other public spaces. Many companies have been successfully deploying computer vision for security and surveillance problems, such as Hikvision (`https://www.hikvision.com`), Avigilon (`https://www.avigilon.com`), Verkada (`https://www.verkada.com`), Huawei, Google, Microsoft, and Amazon. Let's delve into one interesting case study with Fujitsu and its interesting use of computer vision to monitor and smooth traffic flows in Montreal.

The city of Montreal struggled with many issues related to traffic flow because of factors such as limited entry and exit points, insufficient road infrastructure, and an inadequate traffic management system. As a solution, Fujitsu proposed a computer vision-based solution for most traffic issues. The system collects data from CCTV cameras, more than 2,500 traffic lights, and other sensors. Then, the system makes a real-time decision to optimize the traffic flow. The computer vision-based solution has reduced travel time, air pollution, and other traffic-related issues. For more details, please refer to *Smoothing traffic flows with AI analysis* (`https://www2.fujitsu.com/global/customer-stories/cs-city-of-montreal-20210701`).

Synthetic data and computer vision – examples from industry

In this section, you will learn about and understand how companies have just started using synthetic data-based computer vision solutions to stand out from competitors and overcome real data limitations and issues.

Neurolabs using synthetic data in retail

According to *Getting Availability Right: Bringing Out-of-Stocks Under Control* (`https://www.oliverwyman.com/content/dam/oliver-wyman/global/en/2014/aug/2012_CEU_POV_Getting%20Availability%20Right_ENG.pdf`), out-of-stock items cause heavy financial losses and dissatisfied customers. The consequences can be dramatic on businesses and revenues. At the same time, collecting and annotating large-scale real data for this task is an expensive and time-consuming process. Neurolabs, an ML company specializing in providing solutions in retail automation, investigated an elegant solution using synthetic data for this issue. They utilized Unity and their own synthetic data generator to generate 1,200 images of 129 unique **Stock Keeping Units (SKUs)** on shelves. The dataset is named CPGDet-129 and can be downloaded from this link (`https://dl.orangedox.com/SampleRetailSyntheticDataset`). Additionally, for more details about the dataset and license, please refer to the GitHub repository (`https://github.com/neurolaboratories/reshelf-detection`). The dataset was automatically generated and labeled. Moreover, it specifically supports object detection tasks. Training a state-of-the-art object detection algorithm on their synthetic dataset alone, without using any real data, they were able to achieve 60% **Mean Average Precision (mAP)** on a real test dataset. mAP is a metric used to tell us how accurate the object detection model is at predicting the bounding boxes around the objects of interest. Higher values of the mAP score indicate that our model is making accurate predictions.

This is a perfect example showing how synthetic data can be deployed to solve complex computer vision problems in practice.

For more details, please refer to `https://neurolabs.medium.com/using-neurolabs-retail-specific-synthetic-dataset-in-production-bbfdd3c653d5` and `https://www.neurolabs.ai/post/using-neurolabs-retail-specific-synthetic-dataset-in-production`.

Microsoft using synthetic data alone for face analysis

Face analysis such as face parsing and landmark localization is fundamental for modern industry. The applications range from security and advertising to medical diagnosis. Using synthetic data for face analysis seems inescapable as annotating real images for these tasks not only is extremely hard but also brings ethical and privacy issues. You can refer to *Chapter 3*, where we discussed these issues.

Microsoft is one of the pioneer companies in face recognition technologies. They have many years of research and development in this area. *Face API* is just one example (`https://azure.microsoft.com/en-gb/products/cognitive-services/face`). Their recent work, titled *Fake it till you make it: face analysis in the wild using synthetic data alone* (`https://openaccess.thecvf.com/content/ICCV2021/papers/Wood_Fake_It_Till_You_Make_It_Face_Analysis_in_the_ICCV_2021_paper.pdf`), is an excellent demonstration of how synthetic data can be deployed in computer vision.

The researchers in this work first procedurally generated photorealistic synthetic faces. They used a template face and then randomized the hair, clothes, expression, and, essentially, identity. They simulated these faces in random environments. The synthetic dataset they have generated contains 100,000 synthetic faces with ground-truth annotations, including 2D dense face landmarks and per-pixel face parts semantic segmentation.

They trained face parsing and landmark localization ML models on their generated synthetic data alone without using any real images. Their experimental results show that the trained ML models were able to achieve superior results on real datasets. For example, their synthetic-data-trained ML model was able to predict 10 times more landmarks as compared to real-data-trained ML ones. They claim that this success is due to the superiority of their synthetic training data. They emphasize that it is impossible for human annotators to accurately label that many landmarks in practice. Additionally, they show that their synthetic data generation pipeline can be easily adapted to generate synthetic training data for other computer vision tasks, such as eye-tracking. They simply add a virtual eye-tracking camera and generate training images with the corresponding ground truth. To download the dataset, you can refer to their GitHub repository (`https://microsoft.github.io/FaceSynthetics`).

Synthesis AI using synthetic data for virtual try-on

Virtual fashion is gaining more momentum because it provides a sustainable solution that, unlike traditional fashion, reduces cost and effort. Additionally, it provides a scalable and more personalized solution for companies and customers. Furthermore, it opens more opportunities for creativity, collaboration, and social impact. For this industry to flourish and achieve the intended outcomes, computer vision technologies need to excel at a number of tasks, such as pose estimation, semantic segmentation, visual object detection, and tracking. *Synthesis AI* proposed an elegant solution by using synthetic photorealistic 3D humans with huge variations in body type, gender, ethnicity, age, height, and other attributes. They were able to generate depth maps, surface normals, segmentation maps, and many other annotations. For more details, please refer to *Synthesis AI Virtual Try-on* (`https://synthesis.ai/applications/virtual-try-on`). Additionally, they experimentally demonstrated the usability of the generated synthetic data for a number of tasks, such as face segmentation, background matting, and facial landmark detection. They found that fine-tuning on real data after pretraining on synthetic data achieves the best results as compared to training on real data or synthetic data alone or even a mixture of both. To explore the case study in more detail, please refer to *Synthetic Data Case Studies: It Just Works* (`https://synthesis.ai/2021/06/17/synthetic-data-case-studies-it-just-works`).

Summary

In summary, you were introduced to various industrial applications of computer vision. You learned why and how computer vision is shaping the future of our modern industry. Moreover, you explored two case studies of companies that started to utilize synthetic data for their computer vision-based solutions.

In the next chapter, you will delve into another set of interesting case studies in the field of natural language processing.

11
Case Study 2 – Natural Language Processing

This chapter introduces you to **natural language processing** (**NLP**), where synthetic data is a key player. You will explore various applications of NLP models. Additionally, you will learn why these models usually require large-scale training datasets to converge and perform well in practice. At the same time, you will comprehend why synthetic data is the future of NLP. The discussion will be supported by a practical, hands-on example, as well as many interesting case studies from research and industry fields.

In this chapter, we're going to cover the following main topics:

- A brief introduction to NLP
- The need for large-scale training datasets in NLP
- Hands-on practical example with ChatGPT
- Synthetic data as a solution for NLP problems

A brief introduction to NLP

NLP is an interdisciplinary field that combines computer science, ML, and linguistics. It gives computers the ability to understand, analyze, and respond to natural language texts, written or spoken. The field of NLP is evolving for many reasons, including the availability of big data and powerful computational resources such as **Graphics Processing Units** (**GPUs**) and **Tensor Processing Units** (**TPUs**). Examples of state-of-the-art NLP models include *BERT: Pre-training of Deep Bidirectional Transformers for Language Understanding* (https://arxiv.org/abs/1810.04805), *ChatGPT* (https://openai.com/blog/chatgpt), and *Google Bard* (https://bard.google.com). Next, let's explore some of the key applications of NLP models in practice.

Applications of NLP in practice

Some common applications of NLP models are shown in *Figure 11.1*.

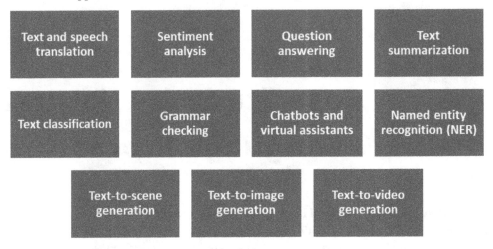

Figure 11.1 – Samples of key applications of NLP models in practice

Let's now discuss some of these applications in more detail.

Text and speech translation

This is the task of translating text or a speech from one language to another. Usually, a large-scale text corpus, composed of a huge number of sentences translated from one language to another, is used to train such models. *Google Translate* (https://translate.google.co.uk), *Microsoft Translator* (https://translator.microsoft.com), and *iTranslate* (https://itranslate.com) are all examples of generic translation NLP models. There are also domain- or field-specific NLP-based translators, such as *Lingua Custodia* (https://www.linguacustodia.finance) and *Trados* (https://www.trados.com), which are more specific to the financial field.

Sentiment analysis

This is a major task in the NLP field. It aims at analyzing and classifying texts based on the sentiment or emotions embedded in the text. It is usually used by companies to understand customer feedback, assess their services, and identify issues. For example, it is commonly employed to classify customer reviews on items or services as *positive*, *negative*, or *neutral*. Additionally, it is often applied to identify emotions in text, such as anger, sadness, dissatisfaction, frustration, and happiness. For example, *Medallia Text Analytics* utilizes NLP to provide a quick summary of market trends and customer feedback and comments on services and products. For more details, please refer to the Medallia Text Analytics website (https://www.medallia.com/resource/text-analytics-solution-brochure). Furthermore, for recent examples of using sentiment analysis in practice, please refer to *Identification of opinion trends using sentiment analysis of airlines passengers' reviews* (`https://doi.org/10.1016/j.jairtraman.2022.102232`) and *A Novel Approach for Sentiment Analysis and Opinion Mining on Social Media Tweets* (`https://link.springer.com/chapter/10.1007/978-981-19-2358-6_15`).

Text summarization

This is the task of generating a summary of text or a document in a human-like way by capturing the essence or the main points. It is a complex process as the NLP model needs to learn how to focus on the essential parts of the text, which is a context-dependent task. However, NLP models have shown great progress in this area recently. There are many examples of NLP models that can summarize large blocks of text, such as the *plnia Text Summarization API* (`https://www.plnia.com/products/text-summarization-api`) and *NLP Cloud's Summarization API* (`https://nlpcloud.com/nlp-text-summarization-api.html`).

Test-to-scene generation

This is another essential task that relies extensively on NLP. It aims at generating virtual scenes given descriptive textual input. It has many interesting applications in game development, the metaverse, advertising, and education. One of the main advantages of text-to-scene generation is that it allows users to generate diverse and photorealistic scenes without requiring a background in computer graphics, game development, and programming. Text-to-scene methods are usually based on GANs, VAEs, diffusion models, and Transformers. For more information, please refer to *Text2NeRF: Text-Driven 3D Scene Generation with Neural Radiance Fields* (`https://arxiv.org/pdf/2305.11588.pdf`) and *SceneSeer: 3D Scene Design with Natural Language* (`https://arxiv.org/pdf/1703.00050.pdf`).

Text-to-image generation

In this task, NLP models generate images based on textual descriptions provided by the user. In this task, the model aims at creating a visual representation controlled by the textual input. The task of generating images from text has many attractive applications, such as data augmentation, content generation, e-commerce, and advertising. You can take *DALL-E2* (`https://openai.com/product/dall-e-2`) and *Stable Diffusion* (`https://stablediffusionweb.com`) as examples. They can generate photo-realistic images given a descriptive text.

In the next section, we will learn why we need large-scale datasets to successfully train NLP models.

The need for large-scale training datasets in NLP

NLP models require large-scale training datasets to perform well in practice. In this section, you will understand why NLP models need a substantial amount of training data to converge.

ML models in general required a huge number of training samples to cover in practice. NLP models require even more training data compared to other ML fields. There are many reasons for that. Next, let's discuss the main ones, which are as follows:

- Human language complexity
- Contextual dependence
- Generalization

Human language complexity

Recent research shows that a huge proportion of our brains is used for language understanding. At the same time, it is still a research problem to understand how different brain regions communicate with each other while reading, writing, or carrying out other language-related activities. For more information, please refer to *A review and synthesis of the first 20years of PET and fMRI studies of heard speech, spoken language and reading* (`https://doi.org/10.1016/j.neuroimage.2012.04.062`). Additionally, infants' basic speech and vision functionalities are developed by 8 to 12 months of age. However, it takes them a few years to use verbal or textual communication appropriately. Thus, language processing is not only hard for computers but also for humans. What makes the problem much harder for machines is the need to learn grammar, expressions, and metaphors. Thus, NLP models require substantial training data to learn these hidden rules and patterns.

Contextual dependence

While most ML tasks are still context dependent, such as computer vision, the contextual dependence is more severe and intense with NLP problems. For example, the meaning of a sentence can change from declarative to interrogative based on the speaker's tone of voice. The ordering of the words and the previous and next few sentences may also change the meaning and imply different interpretations. Even the same words sometimes have different meanings based on the context. For example, "light" can be used as a noun to mean "illumination" or an adjective to mean "little weight." Thus, to master these scenarios, the NLP models need to be trained on a diverse dataset that includes enough examples to cover these contexts.

Generalization

Another issue that makes the NLP task even harder is the huge number of languages, dialects, and varieties. For instance, British English is different from American English in some ways. It is not just different in the spelling but also in the words, idioms, and expressions used. Even within the same country with the same language, people in different cities use different dialects. For NLP models to be useful and practical, they should accommodate most of these variations. Thus, large-scale training data is usually required to capture these variations.

In the next section, we will learn how to generate synthetic data for a practical example in NLP.

Hands-on practical example with ChatGPT

Let's imagine we want to train an NLP model to identify positive and negative reviews of a book about ML. We will use **ChatGPT** (`https://chat.openai.com`) to generate a list of positive and negative reviews. Then, we will deploy these reviews to augment our real training dataset.

To get 10 positive reviews, we can simply use the following prompt: `Give me 10 good reviews about a book for Machine Learning`. ChatGPT will generate the following reviews:

- Mastering Machine Learning is an exceptional book for both beginners and seasoned professionals. The author's clear and concise explanations make complex concepts easy to grasp, ensuring a solid foundation for anyone diving into the world of ML.

- This book strikes the perfect balance between theory and practical implementation. The hands-on exercises and real-world examples provide valuable insights, making it an invaluable resource for anyone looking to apply ML in their projects.

- I appreciate how the author covers a wide range of ML algorithms and techniques, going beyond the usual suspects. It's a well-rounded guide that truly equips readers with the knowledge needed to tackle diverse ML challenges.

- The author's writing style is engaging and approachable, making it feel like a friendly mentor guiding you through the ML journey. Complex topics are broken down into manageable pieces, making the learning process enjoyable.

- The book's emphasis on understanding the math behind ML algorithms sets it apart from others. It builds a solid mathematical foundation without overwhelming readers, which is crucial for aspiring ML practitioners.

- The book's code samples are well-documented and easily reproducible. The practical exercises are a fantastic way to reinforce learning and experiment with ML concepts in a hands-on manner.

- One of the highlights of this book is its practical advice on model selection and hyperparameter tuning. These insights save time and help avoid common pitfalls in ML projects.

- As someone with a non-technical background, I found this book extremely accessible. The author takes the time to explain even the most complex concepts in a manner that anyone can understand.

- The book stays up-to-date with the latest trends and advancements in the field of ML. It includes coverage of recent breakthroughs and contemporary applications, making it a relevant and forward-looking resource.

- This book is a true gem for aspiring data scientists and ML enthusiasts. It not only teaches ML techniques but also imparts a problem-solving mindset, empowering readers to tackle real-world data challenges with confidence.

Similarly, to get negative reviews, we may use the `Give me 10 negative reviews about a book for Machine Learning` prompt. We get the following negative reviews:

- This book was a complete disappointment. The explanations were convoluted, and the author seemed more interested in showcasing their knowledge than helping readers understand the concepts. I felt lost throughout the entire reading experience.

- The book claims to be for beginners, but it assumes a level of prior knowledge that I just didn't have. The lack of clear explanations and practical examples made it nearly impossible for me to follow along.

- I found numerous errors in the code samples provided. It was frustrating to spend hours trying to debug the examples when I was supposed to be learning. Sloppy editing and poor code quality were major drawbacks.

- The book's content felt outdated, given the rapidly evolving field of ML. I expected to learn about cutting-edge techniques, but instead, I got a rehash of old material that's easily available online.

- The book had a promising start, but it quickly became overly technical and hard to digest. The author assumed readers had a PhD in mathematics, making it inaccessible to anyone without an advanced background.

- I was excited to see practical exercises, but the ones in this book were poorly explained and poorly executed. The lack of step-by-step guidance made it impossible to apply the concepts effectively.

- The book's organization was chaotic, jumping between topics without any clear progression. It left me feeling confused and struggling to connect the dots between chapters.

- The examples provided were unrealistic and didn't reflect real-world scenarios. I had trouble relating the concepts to practical applications, which defeated the purpose of learning ML in the first place.

- The book promised to be comprehensive, but it lacked depth in crucial areas like neural networks and deep learning. I was hoping for a well-rounded understanding, but instead, I found huge gaps in the content.

- The writing style was dry and monotonous, making it difficult to maintain interest in the subject matter. I found myself dozing off during what should have been exciting topics in the world of Machine Learning.

These positive and negative synthetic reviews can be used to augment our training dataset. As you can see, it is straightforward to generate an unlimited number of training samples to augment and complement our training data. The generated samples are diverse, relevant, accurate, detailed, and structured. Thus, including such samples in our training dataset can improve the performance of our NLP model.

Next, let's delve into three practical examples of using NLP models in industry.

Synthetic data as a solution for NLP problems

In this section, you will understand how companies are leveraging synthetic data as a solution for their NLP-based problems. We will look at four case studies:

- SYSTRAN Soft's use of synthetic data
- Telefónica's use of synthetic data
- Clinical text mining utilizing synthetic data
- The Alexa virtual assistant model

SYSTRAN Soft's use of synthetic data

Neural Machine Translation (**NMT**) is a promising approach in NLP. It utilizes neural networks to learn statistical models and thus perform the translation task. The typical architecture is composed of an encoder-decoder, which is usually trained on large-scale training datasets. These models were shown to achieve excellent results in practice. However, they also have some limitations, as we will see with the SYSTRAN case study.

SYSTRAN is one of the few pioneering companies in the field of machine translation technology (`https://www.systransoft.com`). While their standard and traditional NLP models achieved state-of-the-art results, they struggled under two main scenarios: translating long sentences and translating short titles, such as titles of news articles. To solve these issues, they explored augmenting their real training data with synthetic data specially generated for that aim. They were able to solve these issues and boost the overall performance. For more information, please refer to *SYSTRAN's Pure Neural Machine Translation Systems* (`https://blog.systransoft.com/wp-content/uploads/2016/10/SystranNMTReport.pdf`).

Telefónica's use of synthetic data

In telecommunication industries, it is essential to collect data about customers to analyze their needs, identify issues, and customize the provided services. This helps these companies to establish a stronger reputation and thus be more successful in the market. The issue is usually not data availability but the regulations that limit utilizing customers' data to train NLP or ML models in general.

Telefónica deployed an elegant solution to address these issues. They used the *MOSTLY AI* synthetic data platform to synthesize a new dataset from the original customer dataset *Telefónica's CRM Datamart*. The newly generated synthetic data now meets the requirements of GDPR as it does not contain any real information about customers. At the same time, the synthetic dataset has patterns, correlations, and statistical properties that can be seen in the original real dataset. Thus, it can be used as a replica of the real dataset to train NLP models. This allowed the company to use up to 85% of the customer data, which was not possible with real data-based NLP models.

Clinical text mining utilizing synthetic data

A recent study conducted by researchers at Rice University and Texas A&M University, as well as other collaborators, investigated the usability of synthetic data generation models such as ChatGPT on clinical text mining. Their aim was to use **Large Language Models** (**LLMs**) to help with clinical texting mining. They deployed LLMs to recognize biological named entities from unstructured healthcare textual data. They interestingly found that using ChatGPT, which was directly trained on real data for this task, did not achieve a satisfactory performance. Developing a synthetic data generation pipeline and generating the necessary synthetic data dramatically improved the performance of their models. The F1-score increased from 23.37% to 63.99%, which is a significant increase. Additionally, they highlighted that their synthetic-data-based model now better addresses and mitigates privacy concerns compared to the real-data-based one. For more information, please refer to *Does Synthetic Data Generation of LLMs Help Clinical Text Mining?* (https://arxiv.org/pdf/2303.04360.pdf).

The Alexa virtual assistant model

Virtual assistant models, such as Alexa by Amazon, Siri by Apple, and Google Assistant by Google, are becoming an integral part of our modern lives. They provide enormous services, such as ordering products, controlling home appliances, and voice searching. For these tools to become beneficial for a wider audience, they need to support many languages and dialects, which requires large-scale training datasets.

One of the main issues the Alexa virtual assistant encountered when Amazon launched the model for three new languages, Hindi, US Spanish, and Brazilian Portuguese, was the scarcity of real training data. As a solution, Amazon leveraged the available limited real data to create "templates." Then, they deployed these templates to generate synthetic data that augmented and complemented the real data. For example, they utilized the available real data in these languages to learn the essential grammar and syntax of the languages. Then, they leveraged the trained models to generate a sufficiently large synthetic training dataset, which consisted of novel sentences following the grammar and syntax of these languages. This elegant synthetic-data-based solution helped Amazon to mitigate real data insufficiency and thus helped the company to provide more accurate virtual assistants for a broader audience with even better performance. Consequently, Amazon successfully got more orders and higher profitability. For more information, please refer to *Tools for generating synthetic data helped bootstrap Alexa's new-language releases* (`https://www.amazon.science/blog/tools-for-generating-synthetic-data-helped-bootstrap-alexas-new-language-releases`).

Summary

In this chapter, we introduced NLP models and explored the main applications of these models in practice. Additionally, we learned that NLP models require large-scale datasets. Then, we thoroughly discussed the main reasons for that. Following this, we studied a few examples from industry and research where synthetic data was successfully deployed. In the next chapter, we will delve into another set of interesting case studies where synthetic data has been successfully deployed in the predictive analytics field.

12
Case Study 3 – Predictive Analytics

This chapter introduces you to predictive analytics, which is yet another area where synthetic data has been used recently. Furthermore, you will explore the key disadvantages of real data-based solutions. The discussion will be enriched by providing examples from the industry. Following this, you will comprehend and clearly identify the benefits of employing synthetic data in the predictive analytics domain.

In this chapter, we will cover the following main topics:

- What is predictive analytics?
- Predictive analytics issues with real data
- Case studies of utilizing synthetic data for predictive analytics

What is predictive analytics?

In this section, you will be introduced to predictive analytics. You will understand its various applications in practice. You will also explore its wide application in different fields, including banking, finance, and healthcare.

Predicting the future has always been a central concern and has captivated humankind for thousands of years, and it's still a hot topic! Predicting the future helps to identify and exploit opportunities, optimize resource allocation, and prepare better for catastrophes, disasters, and crises. Predictive analytics is a subfield of data analysis that aims to utilize historical data to predict or forecast the future. Now, to embark on our journey of understanding predictive analytics, let's examine some of its interesting applications.

Applications of predictive analytics

There are enormous applications of predictive analytics in practice. A sample of these applications is shown in *Figure 12.1*.

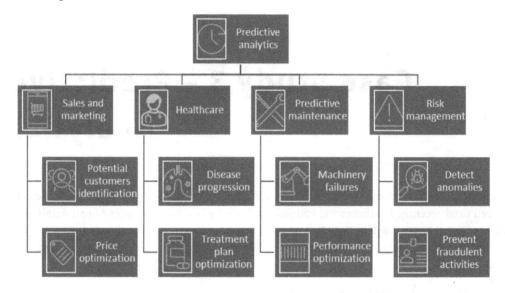

Figure 12.1 – A sample of predictive analytics applications in industries

Let's briefly discuss some of these applications.

Sales and marketing

Identifying potential customers is paramount for companies, as it enables them to concentrate their resources and attention and allocate their advertising efforts effectively toward a specific customer segment. Additionally, it helps companies to create personalized advertisements, products, and services. Thus, they can attract more customers, optimize their resources, and consequently achieve higher profits. For example, *Starbucks* leverages your purchase history to suggest new drinks that you may like. For more information, refer to *How Starbucks Uses Predictive Analytics and Your Loyalty Card Data* (https://demand-planning.com/2018/05/29/how-starbucks-uses-predictive-analytics-and-your-loyalty-card-data).

Moreover, price optimization is another central element for businesses. Companies usually aim to find the ideal price that will achieve the highest profitability. There is often a trade-off between low prices but more customers and high prices but fewer customers. Thus, companies utilize predictive analytics and ML to take into consideration market situations, such as demand, competitors' prices, and item qualities and quantities for optimal price predictions. For more information about pricing optimization, refer to *Pricing Optimization & Machine Learning Techniques* (https://vitalflux.com/pricing-optimization-machine-learning-techniques).

Healthcare

Predicting accurately the progression of a disease helps to control its progression or even prevent it. For example, predicting the progression of cancer is crucial to plan the most effective treatment for patients. In general, chronic diseases especially benefit from utilizing predictive analytics. Researchers have shown that utilizing test results, clinical visits, and other information about the patient can help them to accurately predict chronic disease progression. For more details, read *Chronic Disease Progression Prediction: Leveraging Case-Based Reasoning and Big Data Analytics* (`https://doi.org/10.1111/poms.13532`).

Predictive maintenance

Unplanned downtime is one of the main issues that negatively impact customers, services, a company's reputation, and revenue. Thus, predicting machinery failures can help to avoid them and enhance machine and equipment lifespan. Predictive analytics usually leverage data collected from various sources, including the following:

- Data from sensors, such as pressure, temperature, and vibration frequency
- Data from failure recorders, such as the failure types, the reasons, and the consequences
- Machine historical performance data, such as data collected under standard and non-standard operating conditions
- Industry benchmarks

For more details about predictive analytics applications for equipment failure, refer to *Predictive Maintenance: Employing IIoT and Machine Learning to Prevent Equipment Failures* (`https://www.altexsoft.com/blog/predictive-maintenance`).

Risk management

Predictive analytics is widely used for risk management in modern business. The historical data of previous risk incidents and fraudulent activities can be leveraged to train ML models to predict possible attacks, threats, and risks in the future. *PayPal*, *Apple Pay*, and *Amazon Pay* continuously assess customers' behavior to identify any suspicious activities or abnormal patterns. This includes unusual spending patterns, many failed authentication attempts, and rapid account or customer information changes. The ML models are usually trained on customers' relevant data, such as transaction patterns, device information, and customers' locations. For more information, refer to *4 Ways Machine Learning Helps You Detect Payment Fraud* (`https://www.paypal.com/us/brc/article/payment-fraud-detection-machine-learning`).

Now, let's delve into the main issues of real data-based predictive analytics solutions in practice.

Predictive analytics issues with real data

In this section, you will learn about the main issues with real data-based predictive analytics solutions. Mainly, we will discuss the following three issues.

Partial and scarce training data

One of the main requirements of predictive analytics models to work well in practice is the availability of large-scale historical data. In sectors such as healthcare, banking, security, and manufacturing, it is not easy to find such datasets. The main reasons behind that are privacy concerns, regulations, and trade secrets. As we know, without sufficient training datasets, ML algorithms simply cannot work well in practice. Thus, predictive analytics methods based on real data work well only in certain fields where data is available. Therefore, augmenting real data with synthetic data can complement small-sized and incomplete real datasets. Thus, it solves one of the main issues of real datasets in these fields, as we will see in the next section.

Bias

Certainly, another major issue with real datasets is bias. In other words, data is often not evenly distributed. Thus, predictive models become biased, which leads to wrong decisions and predictions. Using synthetic data, you can make your dataset balanced by generating samples for classes or categories with fewer samples. This positively impacts your predictive model performance and, thus, improves the decision-making procedure of your company.

Cost

Although collecting and annotating real data is often an expensive and time-consuming process, companies must update their data continuously to accurately capture and present the real market. Thus, using real data may cost the company more and yet cause ML models to make predictions based on outdated data. Additionally, as mentioned, companies need to continuously collect and annotate real data, which is indeed a cumbersome process. Synthetic data can be generated quickly and efficiently to augment real data. Thus, it presents a cheap and high-quality solution for predictive analytics problems.

Now, let's explore in detail three case studies of utilizing synthetic data in predictive analytics.

Case studies of utilizing synthetic data for predictive analytics

In this section, you will explore the huge opportunities brought by synthetic data to the field of predictive analytics. You will delve into three interesting case studies:

- Provinzial and synthetic data

- Healthcare benefits from synthetic data in predictive analytics

- Amazon fraud transaction prediction using synthetic data

Provinzial and synthetic data

Provinzial, one of the top insurance companies in Germany, always had many issues with the usability of real insurance data because of the many regulations that limit the usability of sensitive data for predictive analytics in a rapidly changing market, such as the insurance market. At the same time, they wanted to improve their services and the offers they make to their customers. To address these issues, they investigated using synthetic data and, indeed, trained their ML models using the generated synthetic data. The predictive analytics synthetic data-based system, called "Next Best Offer," achieved amazing results, which helped the company to better understand and support customers. Leveraging synthetic data, they were able to train their model on a large-scale dataset of 1 million customers, including 380 features. The dataset included sensitive information about customers such as addresses, payment histories, and insurance claims. They emphasized that using synthetic data helped their company to test their hypothesis without violating customers' privacy. Synthetic data reduced the time required to handle privacy issues. They concluded that using synthetic data-based predictive analytics was beneficial and valuable:

> *"We have found it [a synthetic data-based solution] to be a useful solution for our data science team to simplify data access and focus on our data projects, machine learning model optimizations, and testing new ideas."*
>
> *– Dr. Sören Erdweg, data scientist at Provinzial*

For more information, refer to *Synthetic data for predictive analytics in insurance: The case of Provinzial* (https://www.statice.ai/post/synthetic-data-for-predictive-analytics).

Healthcare benefits from synthetic data in predictive analytics

Chimeric Antigen Receptors Cell Therapy (CAR-T) is an effective treatment for specific types of blood cancer. Usually, the decision of which patients are eligible and can better benefit from this treatment is decided by a dedicated panel (see `https://www.england.nhs.uk/cancer/cdf/car-t-therapy`). ML models and predictive analytics are utilized to make accurate predictions for this task. However, with the limited amount of available real data, the task is complex for ML models. As expected, a slight improvement in the accuracy of such models means more lives to save and less cost to spend. Thus, it is still a hot research topic that requires more work.

In this particular case study, the task is a binary classification problem – eligible or ineligible to receive CAR-T treatment. The main issue here, in addition to data scarcity, is an imbalanced data distribution, where one class has much more training samples compared to the other, which causes the ML models to give biased predictions. Using synthetic data to augment real data was an ideal solution to make the training dataset balanced. This is achieved by generating synthetic samples for the class with less training data.

Utilizing synthetic data for this problem improved the performance by at least three percentage points, and an estimated £6.5 million was saved. Additionally, the solution did not breach any regulations or violate any patients' privacy. Above all of these outcomes, more appropriate patients were identified and, thus, more lives were saved! For more information about this case study, refer to *Predictive analytics in healthcare benefits from synthetic data generation* (`https://mostly.ai/case-study/predictive-analytics-in-healthcare`).

Amazon fraud transaction prediction using synthetic data

According to *UK Finance's Annual Fraud Report of 2023*, in the UK, an estimated £1.2 billion was lost because of various fraudulent activities (`https://www.ukfinance.org.uk/policy-and-guidance/reports-and-publications/annual-fraud-report-2023`). Unfortunately, this is not the only problem; fraudulent activities also hinder economic growth, disrupt relationships, and weaken social cohesion.

Amazon had many issues with the available real datasets that can be leveraged for predictive analytics specifically for fraud transactions. In addition to bias and imbalance issues with real datasets, they did not include sufficient training samples under rare events. This caused ML models to be less accurate in these scenarios. To address these issues, they utilized **WGAN-GP**, which we discussed in *Chapter 7*. They leveraged WGAN-GP to learn real data. Then, they used this generative model to generate samples to make the dataset balanced and include rare fraudulent activities. This supplemented the training dataset and directly improved the robustness and generalizability of the ML model under standard and challenging scenarios, where no sufficient real training data was available. For more information, refer to *Augment fraud transactions using synthetic data in Amazon SageMaker* (`https://aws.amazon.com/blogs/machine-learning/augment-fraud-transactions-using-synthetic-data-in-amazon-sagemaker`).

Summary

In this chapter, we discussed predictive analytics and its major applications in industries. We explored the main limitations of real data-based solutions. Then, we learned how synthetic data presents an elegant solution for these issues. Furthermore, we covered three case studies and learned how some companies have leveraged synthetic data to improve their predictive analytics and, thus, improve their services and revenues. In the next chapter, we will highlight some of the best practices to apply synthetic data successfully and efficiently for your own problems.

13
Best Practices for Applying Synthetic Data

Synthetic data indeed has many advantages and has been successfully and extensively utilized recently in various domains and applications. However, many general issues limit the usability of synthetic data. In this chapter, you will learn about these issues that present a bottleneck for synthetic data. Then, we will delve into domain-related issues that make deploying synthetic data even more challenging. You will explore these issues in various fields, such as healthcare, finance, and self-driving cars. Following this, you will be introduced to an excellent set of good practices that improve the usability of synthetic data in practice.

In this chapter, we're going to cover the following main topics:

- Unveiling the challenges of generating and utilizing synthetic data
- Domain-specific issues limiting the usability of synthetic data
- Best practices for the effective utilization of synthetic data

Unveiling the challenges of generating and utilizing synthetic data

In this section, you will understand the main common issues usually seen across different domains that limit the benefits and usability of synthetic data.

We can roughly categorize these limiting factors into four main categories:

- Domain gap
- Data representation
- Privacy, security, and validation
- Trust and credibility

They can be represented as shown in *Figure 13.1*:

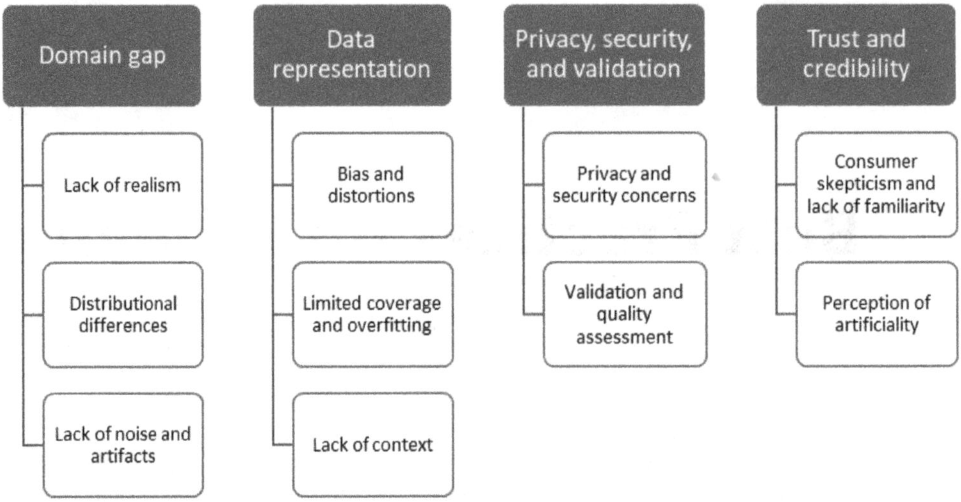

Figure 13.1 – Main factors that limit the usability of synthetic data in practice

Next, let's delve into each of these categories in more detail.

Domain gap

While neural networks are very successful at learning hidden patterns, correlations, and structures in large datasets, they can suffer from the domain gap problem. **Domain gap** usually refers to the difference between the source and target domains' data. The source domain refers to the training data's domain on which the ML model was trained. On the other hand, the target domain refers to the domain on which the model will be tested, evaluated, or used in practice.

In many scenarios, you may achieve excellent results on one dataset but dramatically unsatisfactory performance on another dataset. Both datasets can be real, synthetic, or a mixture of both. However, we focus here on the synthetic source domain and real target domain as it is usually the main and most frequent setup. Training your **Machine Learning** (**ML**) models on a large-scale synthetic dataset and achieving excellent performance in the synthetic domain may not necessarily guarantee the same performance on a real dataset. Thus, it is usually recommended to validate your ML model on a dataset collected from your target domain.

Now, let's dig deeper into the main reasons behind the domain gap between synthetic and real domains. In general, we can identify the following principal factors:

- Lack of realism
- Distributional differences
- Lack of noise and artifacts

Lack of realism

Using a simulator, game engine, or generative model may generate appealing and semirealistic synthetic data but not exactly realistic data. Synthetic data generators cannot capture all the details of the complex real world. It is not necessary for ML models to be trained on data that captures all real-world nuances. However, it needs to be trained on synthetic data that captures and reflects the essential and auxiliary task-relevant details. It is crucial to acknowledge the substantial disparity between both scenarios. Additionally, identifying what is relevant to your ML task may not be a straightforward process. Thus, in general, if you do not train your ML model on sufficiently realistic data, you will end up with a domain gap between your training data and the real world. Thus, your model may not perform well in practice.

Distributional differences

The essence of the training process of **deep learning** (**DL**) and ML models is learning how to make associations between input features and output labels. In other words, for a classification problem, the ML model learns to link certain patterns in pixels colors and locations with the target class label. Thus, when the ML model receives a new image, it may correctly classify it based on the patterns that the ML model learned in the training stage. As you can see, there is a clear assumption that data distributions between source and target domains are identical or close to each other. If the synthetic data distribution is not sufficiently close to the real one, this will make the learned associations, patterns, and correlations from synthetic data not applicable to real data.

For demonstration, let us imagine a scenario that captures the main idea, although it may not reflect the exact reality. If you trained your ML model to do a cats-dogs classification task on synthetic labeled images collected from the *The Secret Life of Pets* animated movie, you would not expect your model to perform well on real data because of the distributional differences problem. For example, the dimensions of the cats and dogs, colors, variations, and densities concerning other objects in the scenes in this movie may not match the ones in the real dataset even though they still may look partially realistic. It is essential to recognize that the issue here is not photorealism but distributional differences.

Lack of noise and artifacts

While synthetic data may be generated to approximate or represent real data, it is usually extremely challenging to model noise and artifacts (anomalies and imperfections) of complex real-world data. The real world is full of imperfections, anomalies, noise, and artifacts. This is due to many reasons, such as random event occurrences, interactions and emergence among complex processes, limitations and errors in sensors and measurement procedures, and even errors because of human intervention. Thus, synthetic data may successfully present the central portion of the distribution, but it may fail to capture anomalies, outliers, and artifacts. Therefore, when the ML model does not observe these scenarios in the training process, it will simply fail when it encounters similar situations in the real world.

For more details about how to bridge the gap between synthetic and real domains, please refer to *Chapter 14*.

Let's now explore the next primary category that restricts the usability of synthetic data.

Data representation

Training data is usually collected or generated to be a proxy of the real world. The human element is always present in this process. Thus, to some extent, our decisions, assumptions, and biases are explicitly or implicitly reflected in how we choose to represent the real world for the ML model. Nevertheless, while it is an issue with real data, it is more vital and problematic with synthetic data, as we will see next.

Biases and distortions

As we know, one of the main methods of generating synthetic data is using generative models, which are trained on real data (for more information, please refer to *Chapter 7*). If the generative model is trained on biased and distorted real data, it will also generate biased data. Then, when the ML models are trained on this data, the decisions will also be biased. Thus, it is very important to thoroughly comprehend and focus on the quality and procedures of our decision-making methodology and assumptions as we generate synthetic data.

Limited coverage and overfitting

The second main issue under this category is synthetic data diversity. Imagine you want to build a 3D virtual world using a game engine to generate synthetic data for a semantic segmentation problem. For your synthetic data to be useful, you need to diversify scene elements such as 3D models, materials, textures, lighting, and camera parameters. Otherwise, your ML model will overfit to a few variations and will fail to generalize well when tested on real data. It should be noted that diversifying these scene elements requires more 3D assets to buy or design, more work and effort, more budget to spend, and more engineers, designers, and programmers.

Lack of context

Unlike real data that is generated by real-world processes, synthetic data is generated artificially by algorithms or systems. Thus, it lacks contextual information that can be essential for learning the phenomenon or task under consideration.

For example, let's say we have created a system to generate synthetic data for the cats-dogs classification problem. Indeed, we can generate thousands of labeled cat and dog images under various attributes, such as lighting conditions, backgrounds, and weather conditions. However, what is vital and much harder to capture with synthetic data is context – in other words, where, when, and how dogs and cats usually appear in the real world. For example, we can usually see them in parks, streets, and residential areas. On the other hand, it is unlikely to see them in hospitals and laboratories. Thus, as you can see, if we are not fully aware of the context of the problem, we may end up generating synthetic training data that lacks context. In this simple scenario, it may be easy to understand the context, but in other scenarios, the context may not be clear and straightforward.

Privacy, security, and validation

As we saw earlier, one of the main issues with real data is privacy (for more information, please refer to *Chapter 3*). Unfortunately, even with synthetic data, it is still a concern.

There is usually a trade-off between data usefulness and privacy when dealing with sensitive data. Synthetic data generators for certain problems in fields such as healthcare and finance are usually trained on sensitive data. Thus, there is a chance that the generated synthetic data may reveal sensitive information.

While generating the synthetic data is the main challenge, there are still other tasks to be performed before deploying synthetic data for your ML problem. Synthetic data needs to be evaluated and validated. Therefore, an effective risk assessment procedure should be performed to ensure that synthetic data is anonymous and still represents the phenomenon under consideration. As privacy attacks evolve, synthetic data generation procedures need to be assessed and monitored continuously to ensure that the generated synthetic data does not breach regulations or disclose sensitive information.

Next, let's explore another interesting factor, which is associated with the sociology of customers.

Trust and credibility

ML in general is still a new, emerging field and synthetic data has only been utilized recently. Thus, it requires time for companies, customers, and ML practitioners to understand and trust synthetic data. Let's discuss the main two elements under this category that usually limit the usability of synthetic data in practice.

Consumer skepticism and lack of familiarity

As companies have just started to deploy more synthetic data-based ML solutions, customers have also started to question the usability of this new approach. One of the main reasons behind this is their misunderstanding of or unfamiliarity with synthetic data generation approaches.

Perception of artificiality

Synthetic data is not collected from the real world. Rather, it is generated artificially. Its synthetic nature causes customers to question its usability and genuineness. Thus, they may question and not trust this new data source or any ML solution based on it.

Now that you have understood the key general issues that limit the usability of synthetic data, let's examine a wide range of domain-specific issues commonly seen in certain fields, such as healthcare and finance.

Domain-specific issues limiting the usability of synthetic data

In addition to general issues that limit the usability of synthetic data in practice, there are also domain-specific issues related to that. In this section, we explore these common domain-specific issues limiting the usability of synthetic data. Let's study synthetic data usability issues in the following three fields: healthcare, finance, and autonomous cars.

Healthcare

ML in healthcare requires large-scale training data. Usually, the data is unstructured, comes from different sensors and sources, is longitudinal (data collected over a long period), is highly imbalanced, and contains sensitive information. The illnesses and diseases that patients suffer from are diverse and complex and depend on a multitude of factors, such as genes, geographic location, medical conditions, and occupation. Thus, to generate useful synthetic training data in the healthcare field, domain experts are usually needed to assess the quality of the generated training data and the validity of the assumptions made by ML engineers. For more information, please refer to *Amplifying Domain Expertise in Clinical Data Pipelines* (https://www.ncbi.nlm.nih.gov/pmc/articles/PMC7677017).

Finance

This field is usually associated with rapid changes, being influenced by a huge number of factors and elements that are usually very hard to predict, such as politics, regulations, competitions, new technologies, and natural catastrophes. Thus, it is not easy to generate synthetic data that takes into consideration the dynamics of the market and other factors. Consequently, applying domain knowledge to the synthetic generation pipeline may significantly improve the usability of the generated synthetic data for this field. For more details, please refer to *Expectations, competencies and domain knowledge in data- and machine-driven finance* (https://www.tandfonline.com/doi/full/10.1080/03085147.2023.2216601).

Autonomous cars

Simulating 3D virtual worlds is a hard task. However, what is more challenging is simulating drivers' and pedestrians' behaviors. In the real world, human behaviors are complex, hard to anticipate, and highly dependent on the environment and situation. For example, drivers and pedestrians may not obey traffic regulations and rules in the events of natural disasters and evacuations. Generating synthetic data that incorporates and anticipates similar scenarios is very complex and not easy to achieve. Additionally, simulators usually need to make many assumptions to simplify computations. However, the consequences of these assumptions may not always be clear and may cause ML models to fail in critical and rare situations.

Next, let's learn some best practices to unlock the full potential of synthetic data.

Best practices for the effective utilization of synthetic data

In this section, we will learn about some common good practices that can improve the usability of your synthetic data-based ML solution in practice:

- **Understand the problem**: Before you start deploying synthetic data, you need to understand what the problem with your ML model and data is and why the available real datasets are not suitable. Do not jump directly to the synthetic data solution if you are not fully aware of the problem and the limitations of the available real data-based solutions.

- **Understand the synthetic data generation pipeline**: We should not consider the synthetic data generation pipeline as a black box. However, we need a good understanding of the generation process to avoid biases and artifacts. For example, suppose we are generating synthetic data for an application to flag fraudulent transactions. If our synthetic data generator often generates the majority of fraudulent transactions with certain attributes, such as a transaction amount between 10K and 12K and the transaction location being some specific country, our ML model, trained on this biased data, will tend to mistakenly identify any transaction with these attributes to be fraudulent regardless of other crucial attributes! As expected, this will make our ML model perform poorly in practice.

- **Diversity, variability, and realism**: For synthetic data to be useful in practice, it should usually be diverse, rich, and realistic and match the distribution of real dataset counterparts. Please refer to *Diversity in Machine Learning* (https://arxiv.org/pdf/1807.01477.pdf) and *Enhancing Photorealism Enhancement* (https://arxiv.org/abs/2105.04619). It is always recommended that you analyze the available real data (if any) and identify the key variabilities and properties that you wish your synthetic dataset to address.

- **Continuously validate and evaluate**: You should always and frequently compare and assess the quality of the generated synthetic data to ensure that the data generation pipeline is working as expected. For example, if you are working with sensitive data, you should continuously assess the generated synthetic data to ensure that it does not disclose any sensitive information and to ensure a high-quality anonymization procedure.

- **Combine synthetic with real data**: It is often suggested to combine synthetic with real data to achieve the best results. Training on a mixture of both or pre-training on synthetic data and fine-tuning on real data are well-known approaches to improve the usability of synthetic data. Please refer to *Semantic Segmentation under Adverse Conditions: A Weather and Nighttime-aware Synthetic Data-based Approach* (`https://bmvc2022.mpi-inf.mpg.de/0977.pdf`) and *Using synthetic data for person tracking under adverse weather conditions* (`https://doi.org/10.1016/j.imavis.2021.104187`).

- **Noise and anomalies**: One of the main common issues is ignoring or underestimating the benefits of outliers and rare scenarios when generating synthetic data. Try to always include these circumstances as they are essential to ensure that your ML model does not fail in these situations in the real world.

Summary

In this chapter, we discussed the primary challenges of deploying synthetic data. Then, we delved into domain-specific issues. We learned why synthetic data is inherently challenging, especially in fields such as healthcare and finance. Finally, we explored a list of best practices to improve the usability of your synthetic data in practice. Next, we will focus in more detail on enhancing and improving synthetic data usability through synthetic-to-real domain adaptation techniques.

Part 5:
Current Challenges and Future Perspectives

In this part, you learn about a well-known issue that usually limits the usability of synthetic data. You will learn about the domain gap problem and why diversity and photorealism are some of the main challenges toward generating useful and large-scale synthetic data in practice. You will learn about various approaches to bridge the domain gap and improve the diversity and photorealism of your synthetic data. Then, we will recap the benefits of synthetic data-based solutions, challenges, and limitations. Finally, we will highlight some interesting future perspectives.

This part has the following chapters:

- *Chapter 14, Synthetic-to-Real Domain Adaptation*
- *Chapter 15, Diversity Issues in Synthetic Data*
- *Chapter 16, Photorealism in Computer Vision*
- *Chapter 17, Conclusion*

14
Synthetic-to-Real Domain Adaptation

This chapter introduces you to a well-known issue that usually limits the usability of synthetic data, called the domain gap problem. In this chapter, you will learn various approaches to bridge this gap, which will help you to better leverage synthetic data. At the same time, the chapter discusses current state-of-the-art research on synthetic-to-real domain adaptation. Thus, you will learn which methods you may use for your own problems. Then, it represents the challenges and issues in this context to better comprehend the problem.

In this chapter, we're going to cover the following main topics:

- The domain gap problem in ML

- Approaches for synthetic-to-real domain adaptation

- Synthetic-to-real domain adaptation – issues and challenges

The domain gap problem in ML

In this section, we will understand what the domain gap is and why it is a problem in ML. The domain gap is one of the main issues that limit the usability of synthetic data in practice. It usually refers to the dissimilarity between the distributions and properties of data in two or more domains. It is not just associated with synthetic data. However, it is a common problem in ML. It is very common to notice a degradation in the performance of ML models when tested on similar but slightly different datasets. For more information, please refer to *Who is closer: A computational method for domain gap evaluation* (https://doi.org/10.1016/j.patcog.2021.108293).

The main reasons for the domain gap between datasets can be linked to the following:

- Sensitivity to sensors' variations
- Discrepancy in class and feature distributions
- Concept drift

Let's discuss each of these points in more detail.

Sensitivity to sensors' variations

In computer vision, your ML model may perform well on images captured using certain cameras, setups, and parameters but drastically fail under similar inputs captured with different cameras or using different parameters. For instance, your computer vision model may work well on videos captured from a first-person view but drastically fail on videos captured from a third-person view. Therefore, we can see that even the same task, such as action recognition, is usually studied with an emphasis on the person's viewpoint, as in *First-person Activity Recognition by Modelling Subject-Action Relevance* (`https://doi.org/10.1109/IJCNN55064.2022.9892547`). Another example is the camera's **Field of View (FoV)**. Some computer vision tasks, such as semantic segmentation, are also studied under certain camera FoVs. A semantic segmentation method trained on the Cityscapes and Synscapes datasets, which are captured under a standard FoV, will fail at segmenting fisheye images captured by a super-wide fisheye lens. For an example, please refer to *FPDM: Fisheye Panoptic segmentation dataset for Door Monitoring* (`https://ieeexplore.ieee.org/stamp/stamp.jsp?tp=&arnumber=9959151`).

Discrepancy in class and feature distributions

Discrepancy or inconsistency in the distribution of attributes, classes, and features is one of the main reasons for the domain gap between synthetic and real domains or even in the same real domain when you train and test on different datasets. This can be clearly observed when working on time series-based problems. Many times, the training data becomes outdated and does not come from the same distribution as the test or evaluation data. For example, an ML model trained to predict inflation rates based on data collected one year ago may not perform well once applied to current data because of the domain gap problem. This is because the source and target data distributions and characteristics are now different from expected.

Concept drift

Concept drift refers to the change in the relation between the input (features) and output (target). Let's take an illustrative example. Assume we have designed an object classifier and the "printer" is one of our objects of interest. As you can see in *Figure 14.1*, the shape, color, appearance, and other features of the "printer" concept have drastically changed over time, from early typewriters, to inkjet printers, to laser printers. Thus, if your training data contained only old printers (typewriters), it would simply struggle to accurately classify modern printers because of the domain gap problem.

Figure 14.1 – An example of concept drift (source: Pixabay)

Next, let's explore the main solutions to mitigate the domain gap problem specifically between synthetic and real domains.

Approaches for synthetic-to-real domain adaptation

In this section, you will learn the key approaches for synthetic-to-real domain adaptation. We will discuss the following methods:

- Domain randomization
- Adversarial domain adaptation
- Feature-based domain adaptation

Let's start with one of the most commonly used approaches for domain adaptation.

Domain randomization

Domain randomization is a mechanism or procedure usually used to mitigate the domain gap problem and improve the performance of ML models on the target domain. This approach aims at randomizing the main properties and attributes of the training data or environment, such as simulators to increase the diversity of the scenarios the ML model is exposed to in the training stage. Thus, we can increase the robustness of the ML model for scenarios that it may encounter in the future. For more information, please refer to *Domain Randomization for Transferring Deep Neural Networks from Simulation to the Real World* (https://arxiv.org/pdf/1703.06907.pdf).

Let's examine the main elements that are usually randomized in two interesting fields: computer vision and NLP.

Computer vision

In almost any computer vision problem, task, or system, we have the following four main elements, as shown in *Figure 14.2*:

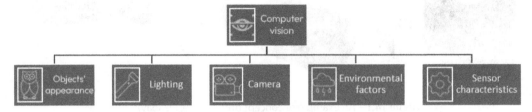

Figure 14.2 – Main elements to randomize in computer vision

Next, let's discuss each of these elements in more detail.

- **Objects**: They are usually the primary focus of computer vision. They populate the scene and interact with each other, the lighting, and the environment. Many major tasks in computer vision, such as object detection, recognition, segmentation, and tracking, are fundamentally related to the appearance of these images in the 3D world and how the world is projected by camera as 2D images.

  Utilizing domain randomization to mitigate the domain gap, we can, for instance, randomize the following elements to diversify the object's appearance:

  - Textures and materials

  - Colors

  - Shapes

  - Dimensions

  - Deformation and animation

  As you can see, deciding which factors are more relevant depends on the task and the problem.

- **Lighting**: Although lighting is necessary to make the objects visible to the observer, it creates a daunting problem for almost all computer vision tasks. A slight variation in lighting conditions drastically changes the appearance of objects, thus changing the pixels' intensity, which makes these objects rather hard for ML-based computer vision models to recognize, detect, or track.

  Thus, to make our ML model robust under various lighting conditions, we may need to diversify the following elements:

  - Lighting intensity

  - Lighting color and temperature

- Light sources

- Lighting anomalies: flares, glares, and flickering

- **Camera and sensor characteristics**: A camera captures visual data from the scene, which is the most generic and conventional input for computer vision models. Thus, to ensure that our ML model generalizes well even under new camera setups, we need to diversify, for instance, the camera setups in the training stage. This can be done by varying the following:

- Camera position

- Orientation

- Altitude

- Viewpoints

- Aperture, exposure, and focus

Additionally, a change in camera parameters such as focal length may also drastically change how the world is captured and perceived. Thus, it will directly affect how the computer vision system recognizes the world, too. Therefore, we may need to consider the following sensor characteristic variations in our training data:

- Lens distortion

- Vignetting

- Chromatic aberration

- Scratches

- Haze

- Light leakage

- **Environmental factors**: Even when the scene, lighting, and camera parameters are the same, environmental factors can substantially change how the scene may look. For example, a weather condition such as fog works as a low-pass filter that removes details of objects that are far from the camera. Thus, it makes extracting robust features for these scenarios harder. Consequently, many ML models may fail or struggle in similar cases.

In many situations, we cannot clearly identify the environmental factors our computer vision system will work under, therefore we may need to randomize the following factors:

- Weather conditions

- Time of day

- Indoor and outdoor environment

- Pollution level

- Wind effects

- Terrain and landscapes

- Road conditions

- Crowd density

- Background clutter

- Geographical locations

Therefore, and based on what we have learned in this section, to mitigate the domain gap between source (training) and target (evaluation) domains, we need to diversify and randomize the scenarios that our ML model will learn during training. Next, let's delve into utilizing domain randomization for NLP problems.

NLP

Similar to what we discussed in computer vision, domain randomization can also be utilized for NLP problems. Usually, it can be deployed to make the NLP models more robust and accurate. For more information, please refer to *Scaling Up and Distilling Down: Language-Guided Robot Skill Acquisition* (`https://arxiv.org/abs/2307.14535`).

As you can see in *Figure 14.3*, there are four key elements that can be randomized to improve the generalizability of NLP models in practice:

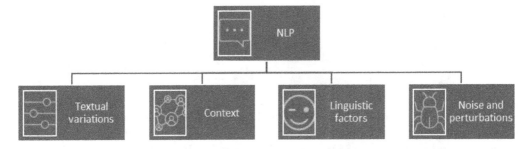

Figure 14.3 – Main elements to randomize in NLP

Let's discuss these elements in detail as follows:

- **Textual variations**: To improve the NLP model's generalizability and robustness to real-world problems, many textual variations are introduced to augment and complement the training data. This is usually done by varying the following elements:

  - Vocabulary

  - Sentence structure

  - Sentence length

- **Context**: NLP models can be utilized in different contexts and for various applications. For example, ChatGPT can be used to answer questions about various topics, such as healthcare, math, finance, and history. It can be utilized to propose travel plans or even to summarize text. Thus, it is crucial to diversify these elements in the training data:

  - Topic

  - Domain

  - Textual genres, such as social media, research papers, and novels

- **Linguistic factors**: An ideal NLP model should be able to handle and respond to queries with different formats and styles. Additionally, it should be capable of understanding sentiments in various applications. Thus, the following factors should be randomized during the training stage to ensure better performance in practice:

  - Style, such as formal, informal, technical, or colloquial

  - Sentiment, such as positive, negative, or neutral expressions

- **Noise and perturbations**: Introducing perturbations in the training data and guiding the NLP model on how to perform these scenarios will ensure that your ML model learns how to respond correctly to these issues. The real world is noisy and thus it is crucial to pay attention to the following factors in the training stage to cover these well-known imperfections usually observed in textual data:

  - Spelling errors

  - Grammar errors

  - Punctuation errors

Next, let's explore another interesting domain adaptation method.

Adversarial domain adaptation

Adversarial domain adaptation is another powerful technique used to bridge the gap between synthetic and real domains based on GANs. In this domain adaptation method, the generator tries to extract domain-independent features while the discriminator tries to identify the source of the data: synthetic or real. Once the model is trained and the discriminator can no longer identify the source of the data domain, the generator can then generate domain-invariant features. For more information, please refer to *Adversarial Discriminative Domain Adaptation* (`https://arxiv.org/pdf/1702.05464.pdf`).

Let's briefly discuss one example from computer vision that illustrates how this synthetic-to-real domain adaptation method can be utilized. For example, let's consider the vehicle re-identification problem. It was shown that utilizing synthetic and real data for training using adversarial domain adaptation improved the performance on two evaluation real datasets, *CityFlow-ReID* (`https://paperswithcode.com/dataset/cityflow`) and *VeRi* (`https://github.com/VehicleReId/VeRi`), with a good margin compared to other solutions. The approach was trained on a mixture of synthetic and real vehicle re-identification datasets. The ML model was guided to learn discriminative features, especially from the real training images. At the same time, it was directed to learn the features that are common between synthetic and real domains. For more details, please refer to *StRDAN: Synthetic-to-Real Domain Adaptation Network for Vehicle Re-Identification* (`https://arxiv.org/abs/2004.12032`). Next, let's look at another approach used for domain adaptation.

Feature-based domain adaptation

Unlike the previous approach, **feature-based domain adaptation** aims at learning a transformation that extracts domain-independent features across synthetic and real domains. This approach transfers the features of both domains into a new representation and then minimizes the discrepancy between the synthetic and real domains. In other words, this method tries to align the feature distributions in both domains with each other. Thus, the approach urges the ML model to learn essential features while discarding domain-specific details and variations. For instance, if the ML model is learning semantic segmentation by being trained on the *Synscapes* (`https://synscapes.on.liu.se`) and *Cityscapes* (`https://www.cityscapes-dataset.com`) datasets, we want the model to learn how to segment humans, cars, traffic lights, and other objects by somehow learning the meaning of these objects. For instance, humans usually walk on pedestrian areas or sidewalks and they usually have a capsule-like shape. These are the sorts of high-level and domain-invariant features that we want the model to learn.

Finally, for a detailed overview of domain adaption methods, please refer to *A Brief Review of Domain Adaptation* (`https://arxiv.org/pdf/2010.03978v1.pdf`). Now, we have learned about some of the key approaches usually utilized for synthetic-to-real domain adaptation. Next, let's delve into their common limitations and how to overcome them in practice.

Synthetic-to-real domain adaptation – issues and challenges

In this section, you will explore the main issues and challenges of synthetic-to-real domain adaptation. This will help you to understand the limitations of this approach. Additionally, it will give you a better insight into how to overcome these issues in your own problem. Therefore, we will focus on the following issues:

- Unseen domain
- Limited real data

- Computational complexity
- Synthetic data limitations
- Multimodal data complexity

Let's discuss them in detail in the following subsections.

Unseen domain

In many cases, the aim is to make sure that your ML model will generalize well to new domains. If we know the domain, domain adaptation methods may work. However, sometimes it is not possible to predict the properties of this new domain. For example, assume you have a computer vision model that works well in Europe but you also want this algorithm to work well in China, Africa, the Middle East, or even on Mars! It is not always possible to have advanced knowledge of the environment or the domain where the ML model will be deployed to make appropriate adaptations.

Limited real data

As we know, real data is scarce and expensive. Thus, supervised domain adaptation methods cannot be easily applied to all ML problems. Additionally, the limited availability of paired data between synthetic and real domains makes the problem even harder. Thus, it is indeed complex and cumbersome to learn the mapping from a synthetic to a real domain.

Computational complexity

Domain adaptation methods are computationally expensive and may require resources, time, experience, substantial budget, and domain experts. Thus, it is not easy, and it is sometimes challenging to train these models, especially adversarial domain adaptation ones.

Synthetic data limitations

Synthetic data has some limitations, especially if the data is not generated appropriately. Synthetic data may lack the diversity and realism of real data. Thus, it makes it even harder for adaptation methods to bridge the gap. Please refer to *Chapter 13*.

Multimodal data complexity

Recent state-of-the-art computer vision systems such *Tesla Vision Autopilot* utilize data captured from different sources, such as cameras, LiDAR sensors, radar sensors, accelerometers, and gyroscopes. However, generating and annotating the real data and matching it with synthetic data is not a straightforward process. For example, it is possible to generate semantic segmentation, optical flow, or other relevant ground truths, but it is very difficult to simulate the behavior of accelerometers and gyroscopes in virtual worlds. Additionally, it is very complex to develop a simulator that provides all these ground truths together. In parallel to that, it is rather hard for domain adaptation to learn how to appropriately adapt the data from one domain to another.

Summary

In this chapter, you learned the essence of the domain gap problem in ML. Additionally, you explored the main solutions to mitigate this problem. We focused on domain randomization in computer vision and NLP. Then, you learned about the main issues and limitations of synthetic-to-real domain adaptation. In the next chapter, we will explore and highlight diversity issues in synthetic data to better comprehend the pros and cons of synthetic data in ML.

15

Diversity Issues in Synthetic Data

This chapter introduces you to a well-known issue in the field of synthetic data, which is generating diverse synthetic datasets. It discusses different approaches to ensure high diversity in large-scale datasets. Then, it highlights some issues and challenges in achieving diversity for synthetic data.

In this chapter, we're going to cover the following main topics:

- The need for diverse data in ML
- Generating diverse synthetic datasets
- Diversity issues in the synthetic data realm

The need for diverse data in ML

As we have discussed and seen in previous chapters, diverse training data improves the generalizability of ML models to new domains and contexts. In fact, diversity helps your ML-based solution to be more accurate and better applicable to real-world scenarios. Additionally, it makes it more robust to noise and anomalies, which are usually unavoidable in practice. For more information, please refer to *Diversity in Machine Learning* (`https://arxiv.org/abs/1807.01477`) and *Performance of Machine Learning Algorithms and Diversity in Data* (`https://doi.org/10.1051/MATECCONF%2F201821004019`).

Next, let's highlight some of the main advantages of using diverse training data in ML. In general, training and validating your ML model on diverse datasets improve the following:

- Transferability
- Problem modeling
- Security

- The process of debugging

- Robustness to anomalies

- Creativity

- Customer satisfaction

Now, let's delve into each of these elements in more detail.

Transferability

When you train your ML model on diverse data covering a variety of scenarios, contexts, and environments, you boost the transferability of your ML solution to other applications and domains. The reason for this is that when the model learns how to deal with more diverse situations in the training stage, it becomes more capable of adapting to new, unseen contexts. For more information, please refer to *Can Data Diversity Enhance Learning Generalization?* (`https://aclanthology.org/2022.coling-1.437`).

Better problem modeling

Diverse training data enables the ML model to look at the problem from different perspectives. For example, let's consider that our ML model is learning a semantic segmentation task. Training the model under adverse weather conditions and with objects of various colors, textures, and shapes will help the model to better learn the mapping from the RGB images to the semantic segmentation ones. Thus, it will significantly enhance the performance as the model has already learned how to capture a wider range of patterns under various variations. These patterns and associations between input features and output labels may not be easily identified given a less diverse training dataset.

Security

Research has recently shown that training your ML model on diverse training data can improve the robustness of your model to adversarial attacks. Thus, your model becomes even more reliable and secure with diverse training data. For example, training your ML model on a diverse set of adversarial training samples significantly boosts your ML model's robustness to adversarial attacks. These attacks primarily involve manipulating images with noise to fool the ML model while still making them recognizable to the human eye. For instance, a slight change in some pixels' intensity such as the color of a traffic sign may cause ML models to wrongly classify it under a different class with a high confidence. For more information, please refer to *Diversity Adversarial Training against Adversarial Attack on Deep Neural Networks* (`http://www.mdpi.com/2073-8994/13/3/428`) and *Adversarial Attacks on Traffic Sign Recognition: A Survey* (`https://arxiv.org/pdf/2307.08278.pdf`).

Process of debugging

Leveraging diverse data in the validation and evaluation stages helps ML practitioners identify the weaknesses and limitations of their ML models and algorithms. Thus, they can avoid costly failures under challenging scenarios. Furthermore, they can iterate on their solutions and mitigate any potential issues and problems. For instance, suppose we are proposing a new person identification ML model. To clearly understand the limitations of our model, we need to evaluate the model on diverse datasets that cover various illumination conditions, camera viewpoints, indoor and outdoor scenes, and other relevant attributes. This will help us to spot the weaknesses of our approach. By returning to our example, we may see that our model is struggling to identify people at nighttime or when the camera is very close to the person. This sort of observation is essential for improving the model and preventing costly failures, which cannot be achieved without using appropriate diverse validation or evaluation datasets.

Robustness to anomalies

Training your ML model on diverse training data that includes anomalies helps the model learn how to deal with similar situations. Thus, it makes your ML-based solution more robust against outliers and unexpected situations. For instance, let's suppose you trained your ML model for depth estimation on standard data collected by an industry-standard sensor under normal conditions. Your model may fail if the camera sensor was partially damaged, or some dust or raindrops accumulated on the camera lens. Therefore, training your ML model on similar scenarios improves the robustness and reliability of your ML system. Please refer to *A Novel Cross-Perturbation for Single Domain Generalization* (https://arxiv.org/pdf/2308.00918.pdf) for more in-depth details.

Creativity

In problems where creativity is a key requirement, training generative models on diverse training data is necessary to fuel this need. For instance, an LLM or image generator will significantly benefit from being trained on textual or visual data collected from different sources. This will help these generative models to be exposed to various topics, styles, ideas, and opinions, which will provide sufficient knowledge and urge the model to be more creative at various tasks and applications. For some interesting examples, please refer to *Deep Dream Generator* (https://deepdreamgenerator.com), *AutoDraw* (https://www.autodraw.com), *Stablecog* (https://stablecog.com), and *DALL-E 2* (https://openai.com/dall-e-2).

Inclusivity

Deploying diverse training data that appropriately represents the real world helps customers to feel that your ML-based solution is inclusive and does not discriminate against any characteristics of the population worldwide. Therefore, it is essential to ensure that your ML model works as intended for all customers regardless of their age, race, gender, geography, language, and religion. If customers feel that they are disadvantaged because of any of the previous factors, they will develop a negative impression of your business, not just the application itself. Additionally, it may cause legal issues and unwanted consequences to organizations. On the other hand, it helps decision-makers to make more appropriate decisions that take into careful consideration the unique needs of each demographic group of the target audience.

Now that we understand the importance of training our ML models on diverse training data, let's examine how to generate diverse synthetic data.

Generating diverse synthetic datasets

In this section, you will learn different methods of generating diverse synthetic datasets. We will discuss the following:

- Latent space variations
- Ensemble synthetic data generation
- Diversity regularization
- Incorporating external knowledge
- Progressive training
- Procedural content generation with game engines

Latent space variations

Latent space usually refers to a high-dimensional space where the training data is represented in a more abstract or compact way. Deep learning with many layers is designed to make the features in the latent space capture more semantic and conceptual information. For more details, please refer to *Chapter 1*. Thus, these features, in that space, convey encoded information about the problem through the ML model during the training stage. We may not be able to directly link the changes in the latent space to the changes that will happen on the generated images in models such as GANs. However, it was shown in *Interpreting the Latent Space of GANs for Semantic Face Editing* (https://arxiv.org/abs/1907.10786) and *Closed-Form Factorization of Latent Semantics in GANs* (https://arxiv.org/abs/2007.06600) that changing certain attributes in the latent space can generate unique and diverse synthetic samples. For instance, if you carefully change certain features in the latent space, you may generate new samples with different poses, backgrounds, lighting, and weather conditions.

Ensemble synthetic data generation

One of the approaches that is usually deployed to improve the diversity of the generated synthetic data is using multiple generative models to ensure that they capture the intended data distribution. This is especially applicable if the distribution is complex and cannot be modeled using a single generative model. For more information, please refer to *Ensembles of GANs for Synthetic Training Data Generation* (`https://arxiv.org/pdf/2104.11797.pdf`). In this work, multiple GANs were used to improve the generated synthetic data diversity. The researchers focused specifically on the effectiveness of this approach for synthesizing digital pathology patches. GANs were trained independently and in isolation from each other on the training dataset. This work shows that the stochasticity of the optimization process is fundamental to better represent the training data distribution and enrich the generated data diversity for GANs.

Diversity regularization

Another approach to encourage generative models to generate diverse synthetic samples is to utilize a regularization term in the training objective or loss. In other words, you can penalize the generative model for generating similar synthetic samples. Thus, your model will tend to generate more diverse samples to minimize the training loss. For example, this approach was utilized in *Mode Seeking Generative Adversarial Networks for Diverse Image Synthesis* (`https://arxiv.org/pdf/1903.05628.pdf`) to address the mode collapse issue in GANs and improve the diversity of the generated synthetic images. For more details about the mode collapse issue in GANs, please refer to *Chapter 7*. This approach does not require any modification to the architecture of the GAN. It simply changes the loss to encourage the generator to generate dissimilar images. Thus, the generator is urged to better cover the training data distribution and consequently generate more diverse synthetic images. For a survey of the regularization approaches in GANs, please refer to *A Systematic Survey of Regularization and Normalization in GANs* (`https://arxiv.org/abs/2008.08930`).

Incorporating external knowledge

You can also condition the generation process to encourage the generative models to generate synthetic data with certain attributes and under specific scenarios. For example, if your data has fewer training samples taken in rainy conditions, you can explicitly condition the GAN model to generate more examples under this weather condition. Additionally, you may prevent generative models from generating examples that are not relevant to your problem. For example, if you are generating cat images in an indoor environment, you may prevent your GAN from generating examples under adverse weather conditions as they are not valid in this particular environment. This can be achieved through various means, such as modifying the loss function to impose penalties on these irrelevant or unwanted predictions. In this scenario, the discriminator would need to make at least two distinct predictions: one for assessing whether the sample is real or fake and another for determining its relevance.

Progressive training

Another interesting approach to increase the diversity of the generated synthetic samples is to gradually introduce more complex patterns, add more layers to the generator and discriminator throughout the training process, and penalize for less diverse examples during the training stage. This encourages the synthetic data generation model to generate more diverse and variant data. For example, researchers in *Progressive Growing of GANs for Improved Quality, Stability, and Variation* (`https://arxiv.org/pdf/1710.10196.pdf`) showed that growing the generator and discriminator by adding new layers and training on more detailed and higher-resolution images as the training progresses significantly improves the stability of the training process of the GAN and the diversity of the generated synthetic images.

Procedural content generation with game engines

Procedural Content Generation (PCG) is a widely used approach in video games to make the virtual world diverse and rich, resulting in a better player experience. The same concept can be utilized in game engines and simulators to create diverse 3D virtual worlds and thus generate diverse synthetic data. PCG can be utilized to generate textures, objects, maps, animations, and other scene elements. For a specific example, please refer to *ProcSy: Procedural Synthetic Dataset Generation Towards Influence Factor Studies Of Semantic Segmentation Networks* (`https://uwaterloo.ca/waterloo-intelligent-systems-engineering-lab/procsy`).

So far, we have learned the main approaches usually utilized to improve the diversity of the generated synthetic data. Next, let's learn the main issues and limitations of these approaches.

Diversity issues in the synthetic data realm

As we have seen, diversity helps us to build robust, accurate, and general-purpose ML models. Additionally, we learned many approaches to improve synthetic data diversity in practice. In this section, we will examine three main issues we usually encounter when we try to generate diverse synthetic data:

- Balancing diversity and realism
- Privacy and confidentiality concerns
- Validation and evaluation challenges

Balancing diversity and realism

There is usually a trade-off between diversity and realism. Generating diverse synthetic examples without considering the realism of these generated samples may introduce or increase the domain gap between synthetic and real domains. For more details, please refer to *Chapters 13* and *14*. For example, let's suppose that we want to generate images with sports cars for a particular computer vision task or application. While it is crucial to generate diverse sports cars that cover most of the available real sports car samples in the real world, we do not want to generate sports cars that are unlikely to be observed in our problem context. Thus, our aim should always be to generate synthetic data that accurately represents the distribution in the real world. It should be diverse but also realistic to be useful for training and testing ML models in practice.

Privacy and confidentiality concerns

When generating synthetic data for applications that have restrictions on the real data because of privacy or confidentiality concerns, it becomes rather hard to generate diverse synthetic data. The reason behind this is the limited understanding of the attributes, patterns, and correlations of the real data, which cannot be learned by generative models given a small-scale training dataset. Thus, it becomes extremely hard for generative models to generate diverse synthetic data for such applications. Please refer to *Chapter 3* for an in-depth discussion about privacy issues with large-scale real datasets.

Validation and evaluation challenges

One of the main issues in this area is assessing the diversity of the generated synthetic data, and thus the usability of this data in practice. Developing a robust, reliable, and universal diversity evaluation metric would be highly beneficial in practice. State-of-the-art metrics are usually problem dependent and experimental and lack the appropriate theoretical framework. For more information, please refer to *Reliable Fidelity and Diversity Metrics for Generative Models* (`https://arxiv.org/abs/2002.09797`).

Summary

In this chapter, we have discussed the main reasons why the diversity of data is crucial for ML-based solutions. We also examined the key approaches to generating diverse synthetic data. Then, we highlighted the main issues and challenges. In the next chapter, we will focus on another relevant and interesting issue in synthetic data, which is photorealism.

16
Photorealism in Computer Vision

In this chapter, you will learn why we need photorealistic synthetic data in computer vision. Then, you will explore the main approaches to generating photorealistic synthetic data. After that, you will comprehend the main challenges and limitations. Although this chapter focuses on computer vision, the discussion can be generalized to other domains and applications.

In this chapter, we're going to cover the following main topics:

- Synthetic data photorealism for computer vision

- Photorealism approaches

- Photorealism evaluation metrics

- Challenges and limitations of photorealistic synthetic data

Synthetic data photorealism for computer vision

In this section, you will learn why photorealism is essential in computer vision. Photorealism of synthetic data is one of the main factors that mitigates the domain gap between real and synthetic data. Thus, training computer vision models on photorealistic synthetic data improves the performance of these models on real data. For more details, please refer to *Hypersim: A Photorealistic Synthetic Dataset for Holistic Indoor Scene Understanding* (https://arxiv.org/abs/2011.02523) and *A Review of Synthetic Image Data and Its Use in Computer Vision* (https://www.ncbi.nlm.nih.gov/pmc/articles/PMC9698631). Additionally, synthetic data can be used to evaluate computer vision algorithms. However, evaluating these models on non-photorealistic synthetic data may cause these models to show poor performance not because of the challenging nature of the test scenarios but because of the domain gap itself. Thus, photorealistic synthetic data is essential to effectively train and accurately evaluate ML models.

Now, let us discuss the main benefits of utilizing photorealistic synthetic data. First, let us delve into feature extraction.

Feature extraction

Computer vision algorithms usually rely on automatic feature extraction, which is learned in the training stage. ML models learn how to identify the most reliable and discriminative features and patterns, which subsequent submodules leverage to learn the actual task, such as semantic segmentation, depth estimation, and visual object tracking. Training your computer vision model on non-photorealistic synthetic data that oversimplifies the real world will lead to inappropriate feature extraction. Conversely, photorealistic synthetic data helps the ML model to learn how to extract discriminative features. Thus, the ML model will perform well in the real world. This is because realistic data helps ML models to better understand the relationship between scene elements, how they affect each other, and how they contribute to the task being learned.

Domain gap

Photorealistic synthetic data mitigates the domain gap between synthetic and real domains. The main reason is that realistic data partially resamples the real data, which helps computer vision models to be trained on data that is closer to the environment where the model will be deployed in practice. Thus, the model can still generalize well from the synthetic data learned in the training stage. On the other hand, large-scale, diverse, but non-realistic synthetic data may enlarge the domain gap and significantly hinder the performance.

Robustness

Creating simulators that simulate realistic lighting, textures, shaders, animations, and camera movements enables researchers to generate large-scale and diverse synthetic training datasets that properly reflect the challenges and varieties in the real world. Thus, computer vision algorithms can be trained on more real scenarios to learn how to adapt to the actual complexities of the real world. This is important to make computer vision algorithms more robust in the real world where collecting real data is extremely expensive or not applicable.

Benchmarking performance

Synthetic data provides a more accurate and efficient way to generate the ground truth (refer to *Chapter 5*). The ground truth is essential for assessing ML models' performance. Photorealistic synthetic data enables us to ensure that the performance of synthetic data appropriately reflects that expected in the real world. Conversely, non-realistic synthetic data is less suitable for accurate evaluation and benchmarking.

Photorealism approaches

In this section, you will learn about and explore the main approaches usually deployed to generate photorealistic synthetic data. We will learn about the following:

- Physically based rendering
- Neural style transfer

Physically Based Rendering (PBR)

The **Physically Based Rendering** (**PBR**) approach is widely used in game engines such as *Unity* and *Unreal* to accurately simulate how materials in the 3D virtual world interact with light. In the real world, this is a complex process, thus it requires a significant understanding of optics and many simplifications to make these processes applicable to game engines. **Physically based materials** are essential to this approach. They resemble how similar materials in the real world interact with light. These materials usually have properties and parameters that are calculated based on real measurements from real-world materials. The properties may include absorption, scattering, and refraction coefficients and parameters. It should be noted that the main principle behind PBR is **energy conservation**, which means that light energy reflected and scattered by a material should not exceed the total incoming or received light energy by this material.

As expected, deploying a photorealistic rendering pipeline will help us to simulate and render more accurate and realistic light behaviors and materials. Thus, we can generate more photorealistic synthetic data.

Neural style transfer

Neural style transfer is a well-known technique that transfers an artistic style from one image to another while preserving the content of the latter. This method can be applied to synthetic datasets to improve their photorealism and thus mitigate the domain gap between synthetic and real data. For example, the **Sim2Real**-style transfer model can be deployed to bridge the gap between synthetic and real data for the task of pose estimation. For more information, please refer to *Sim2Real Instance-Level Style Transfer for 6D Pose Estimation* (https://arxiv.org/abs/2203.02069). Additionally, there are many interesting works that explore how to adapt the *GTA5* synthetic dataset (https://www.v7labs.com/open-datasets/gta5), which was generated from the *Grand Theft Auto V* video game, to the real *Cityscapes* dataset (https://www.cityscapes-dataset.com).

Photorealism evaluation metrics

One of the main issues within this subject matter is quantitatively assessing the photorealism of the generated synthetic images. In this section, we will explore the main metrics usually used. We will explore the following:

- **Structural Similarity Index Measure (SSIM)**
- **Learned Perceptual Image Patch Similarity (LPIPS)**
- Expert evaluation

Structural Similarity Index Measure (SSIM)

SSIM is one of the most widely used metrics to measure the structural similarity between two images. It was first introduced in the paper titled *Image quality assessment: from error visibility to structural similarity* (`https://ieeexplore.ieee.org/document/1284395`). The SSIM metric does not compare individual pixels of the two images. However, it considers a group of pixels assuming that spatially close pixels have inter-dependencies. These dependencies can be linked to the actual structure of objects that were captured and presented by the given images. SSIM specifically focuses on the spatial relationships among pixels, such as edges and textures, to assess how close an image is to a reference one.

Recently, it was shown that SSIM may lead to incorrect or unexpected results when utilized to compare images or when included in the training loss of ML models. For more information, please refer to *Understanding SSIM* (`https://arxiv.org/pdf/2006.13846.pdf`).

Learned Perceptual Image Patch Similarity (LPIPS)

LPIPS measures the distance between images in the feature space by leveraging networks trained for computer vision tasks on large-scale datasets, for example, *VGG* trained on the *ImageNet* dataset. It was found that LPIPS gives more similar results to how humans perceive similarity between images. For more information, please refer to *The Unreasonable Effectiveness of Deep Features as a Perceptual Metric* (`https://arxiv.org/abs/1801.03924`). In this paper, it was found that ML models trained on complex visual tasks learn a rich, general-purpose, and useful visual representation of the world. This knowledge can be leveraged to assess the visual similarity between images in a similar manner to how humans may perceive this similarity.

Expert evaluation

In certain applications, we may need to request a domain expert evaluation of the generated synthetic images. For example, assume your generative model is generating synthetic images of **Computerized Tomography (CT)** scans that will be used later to train a cancer prediction ML model. We can still leverage qualitative metrics, such as SSIM to assess structural similarity with a real data counterpart, **Peak Signal to Noise Ratio (PSNR)** to measure the quality of the generated images, **Fréchet Inception Distance (FID)** to give us an idea about the realism and diversity of the generated samples, and LPIPS to assess the perceptual similarity to real data. However, expert evaluation is still essential for these critical problems. Expert evaluation of the synthetically generated data is essential to verify its validity, quality, and diversity. Most importantly, this evaluation is essential to confirm that synthetic data adheres to ethical standards.

Next, let us discuss some of the main challenges in generating photorealistic synthetic data.

Challenges and limitations of photorealistic synthetic data

In this section, you will explore the main challenges that hinder generating photorealistic synthetic data in practice. We will highlight the following limitations.

Creating hyper-realistic scenes

The real world is complex, diverse, and intricate with details. Scene elements in reality have various shapes, sophisticated dynamics, and highly non-linear interactions. Additionally, our vision and perception of the world are limited and subject to many factors, such as cognitive biases and color perception. Additionally, we may judge photorealism differently based on the context and evaluator. For example, what is more photorealistic, realistic foreground objects and a non-realistic background or the opposite? All these aspects together make generating highly realistic scenes rather hard in practice.

Resources versus photorealism trade-off

Budget, time, skills, and other factors limit the photorealism of the generated synthetic data. As expected, simulating realistic worlds populated with high-poly 3D models and diverse, realistic animations necessitates substantial computational resources. Additionally, employing advanced and complex light- rendering mechanisms, such as ray tracing and PBR, further increases the demand for more processing capabilities and resources. **Ray tracing** is a rendering technique that can simulate the realistic behavior of light and its complex interactions with scene elements. Thus, there is always a trade-off observed between resources and photorealism.

Therefore, it is very important to identify what you mean by photorealism for your particular problem. Additionally, you need to carefully consider which metrics you will deploy to assess the quality and photorealism of the generated synthetic data, taking into account the available resources.

Summary

In this chapter, you learned the main reasons that motivate researchers to strive to achieve high photorealism in generated synthetic data for computer vision problems. You learned about two main approaches usually utilized for that aim. Then, you explored well-known quantitative and qualitative measures deployed to assess the photorealism of the generated synthetic data. Finally, you examined some issues that hinder generating ideal photorealistic synthetic data in practice. In the next and final chapter, we will wrap up and conclude the book.

17
Conclusion

In this chapter, we will summarize what we have learned throughout our journey with synthetic data. We will highlight again the main problems with real-data-based ML solutions. Then, we will recap the benefits of synthetic data-based solutions, challenges, and limitations. Finally, we will highlight some interesting future perspectives.

In this chapter, we will cover the following main topics:

- Real data and its problems
- Synthetic data as a solution
- Real-world case studies
- Challenges and limitations
- Future perspectives

Real data and its problems

In the first part of this book, we learned why ML models require large-scale annotated data (*Chapter 1*). Then, we delved into the main issues with real data, and we thoroughly analyzed the key drawbacks of the annotation process (*Chapter 2*). After that, we discussed and learned why privacy hinders ML progress in certain domains, such as finance and healthcare (*Chapter 3*).

Synthetic data as a solution

Then, we introduced synthetic data as a solution to these complex issues and problems (*Chapters 4* and *5*). Following this, we explored various well-known approaches to generating synthetic data (*Part 3*). We delved into utilizing simulators and game engines to generate synthetic data (*Chapter 6*). Then, we explored GANs in detail (*Chapter 7*). Later, in our exploration of a novel synthetic data generation approach, we uncovered valuable insights about utilizing video games as a source of synthetic data (*Chapter 8*). Finally, we extensively studied diffusion models and explored their diverse applications for synthetic data generation (*Chapter 9*).

Real-world case studies

We discussed and presented the effectiveness of utilizing synthetic data in real-life scenarios, by providing diverse real-world examples of the successful utilization of synthetic data in computer vision (*Chapter 10*), natural language processing (*Chapter 11*), and predictive analytics (*Chapter 12*). Then, we learned about some well-known domain-specific issues that limit the usability of synthetic data and uncovered some best practices to effectively utilize synthetic data in practice (*Chapter 13*).

Challenges and limitations

To ensure that you are aware of common issues that usually hinder the effective utilization of synthetic data, we comprehensively explored synthetic-to-real domain adaptation approaches. We thoroughly studied the domain gap problem in ML and learned about the main approaches for synthetic-to-real domain adaptation (*Chapter 14*). Then, we learned why diverse data is essential in ML and discovered the main strategies to generate diverse synthetic datasets. Following this, we highlighted the main issues and challenges of generating diverse synthetic data (*Chapter 15*). After that, we learned why generating photorealistic data is pivotal in computer vision. We also learned about the main approaches to enhancing photorealism and discussed the essential photorealism evaluation metrics. Then, we covered the challenges and limitations of generating photorealistic synthetic data in practice (*Chapter 16*).

Future perspectives

We have explored together the potential of synthetic data. It is now your turn to unlock new opportunities and possibilities in the field of ML. I highly encourage you to join synthetic data communities, such as the Mostly AI synthetic data community on Discord (`https://mostly.ai/synthetic-data-discord-community`) and the Open Community for the Creation and Use of Synthetic Data in AI (`https://opensynthetics.com`). You can always explore synthetic data generation tools, such as Synthesized (`https://www.synthesized.io`), Gretel.ai (`https://gretel.ai`) and Neurolabs (`https://www.neurolabs.ai`).

In the future, synthetic data is expected to be more diverse, more realistic, cheaper, and easier to generate and utilize. Governments and decision-makers are aware of the huge advantages of synthetic data. Many governments, including the American government, have a clear plan to advance **Privacy-Preserving Data Sharing and Analytics** (**PPDSA**) methods (`https://www.whitehouse.gov/wp-content/uploads/2023/03/National-Strategy-to-Advance-Privacy-Preserving-Data-Sharing-and-Analytics.pdf`). As expected, synthetic data is central to these solutions. Thus, even more research, attention, and funding will be given to synthetic data in the near future.

Summary

In this chapter, we provided a brief overview of what is covered in this book. First, we discussed the need for real data and the main problems usually associated with collecting and annotating large-scale real datasets. Then, we saw that synthetic data presents a clever solution that elegantly mitigates most of these problems and issues. Second, we mastered the main approaches to generating diverse and realistic synthetic data. Third, we explored various case studies and learned about the main issues and limitations of synthetic-data-based ML solutions.

Essentially, you have learned how to overcome real data issues and how to improve your ML model's performance. Moreover, you have mastered the art of meticulously weighing the pros and cons of each synthetic data generation approach. You have also acquired best practices to better leverage synthetic data in practice.

Now, as we approach the end of our learning journey with synthetic data for ML, you are well-equipped with the skills and knowledge to push the boundaries of the ML field. You are ready to deploy synthetic data to your own problems, research, and business to make a significant impact in academia, industry, society, and beyond!

Index

Packtpub.com

Subscribe to our online digital library for full access to over 7,000 books and videos, as well as industry leading tools to help you plan your personal development and advance your career. For more information, please visit our website.

Why subscribe?

- Spend less time learning and more time coding with practical eBooks and Videos from over 4,000 industry professionals

- Improve your learning with Skill Plans built especially for you

- Get a free eBook or video every month

- Fully searchable for easy access to vital information

- Copy and paste, print, and bookmark content

Did you know that Packt offers eBook versions of every book published, with PDF and ePub files available? You can upgrade to the eBook version at Packtpub.com and as a print book customer, you are entitled to a discount on the eBook copy. Get in touch with us at customercare@packtpub.com for more details.

At www.packtpub.com, you can also read a collection of free technical articles, sign up for a range of free newsletters, and receive exclusive discounts and offers on Packt books and eBooks.

Other Books You May Enjoy

If you enjoyed this book, you may be interested in these other books by Packt:

Debugging Machine Learning Models with Python

Ali Madani

ISBN: 978-1-80020-858-2

- Enhance data quality and eliminate data flaws
- Effectively assess and improve the performance of your models
- Develop and optimize deep learning models with PyTorch
- Mitigate biases to ensure fairness
- Understand explainability techniques to improve model qualities
- Use test-driven modeling for data processing and modeling improvement
- Explore techniques to bring reliable models to production
- Discover the benefits of causal and human-in-the-loop modeling

Power BI Machine Learning and OpenAI

Greg Beaumont

ISBN: 978-1-83763-615-0

- Discover best practices for implementing AI and ML capabilities in Power BI along with integration of OpenAI into the solution
- Understand how to integrate OpenAI and cognitive services into Power BI
- Explore how to build a SaaS auto ML model within Power BI
- Gain an understanding of R/Python integration with Power BI
- Enhance data visualizations for ML feature discovery
- Discover how to improve existing solutions and workloads using AI and ML capabilities in Power BI with OpenAI
- Acquire tips and tricks for successfully using AI and ML capabilities in Power BI along with integration of OpenAI into the solution

Packt is searching for authors like you

If you're interested in becoming an author for Packt, please visit authors.packtpub.com and apply today. We have worked with thousands of developers and tech professionals, just like you, to help them share their insight with the global tech community. You can make a general application, apply for a specific hot topic that we are recruiting an author for, or submit your own idea.

Share Your Thoughts

Now you've finished *Synthetic Data for Machine Learning*, we'd love to hear your thoughts! Scan the QR code below to go straight to the Amazon review page for this book and share your feedback or leave a review on the site that you purchased it from.

https://packt.link/r/1-803-24540-9

Your review is important to us and the tech community and will help us make sure we're delivering excellent quality content.

Download a free PDF copy of this book

Thanks for purchasing this book!

Do you like to read on the go but are unable to carry your print books everywhere?

Is your eBook purchase not compatible with the device of your choice?

Don't worry, now with every Packt book you get a DRM-free PDF version of that book at no cost.

Read anywhere, any place, on any device. Search, copy, and paste code from your favorite technical books directly into your application.

The perks don't stop there, you can get exclusive access to discounts, newsletters, and great free content in your inbox daily

Follow these simple steps to get the benefits:

1. Scan the QR code or visit the link below

https://packt.link/free-ebook/9781803245409

2. Submit your proof of purchase
3. That's it! We'll send your free PDF and other benefits to your email directly

www.ingramcontent.com/pod-product-compliance
Lightning Source LLC
Chambersburg PA
CBHW080527060326
40690CB00022B/5049